HOW DOES ASPIRIN FIND A HEADACHE?

Also by David Feldman

Do Elephants Jump?

How Do Astronauts Scratch an Itch?

What Are Hyenas Laughing at, Anyway?

How to Win at Just About Anything

Are Lobsters Ambidextrous?
(formerly published as *When Did Wild Poodles Roam the Earth?*)

Do Penguins Have Knees?

Why Do Dogs Have Wet Noses?

When Do Fish Sleep?

Who Put the Butter in Butterfly?

Why Do Clocks Run Clockwise?

Why Don't Cats Like to Swim?
(formerly published as *Imponderables*)

HOW DOES ASPIRIN FIND A HEADACHE?

An Imponderables® Book

David Feldman

Illustrated by Kassie Schwan

Collins

An Imprint of HarperCollins*Publishers*

A hardcover edition of this book was published in 1993 by HarperCollins Publishers.

First Collins edition published 2005.

The Library of Congress has catalogued the hardcover edition as follows:

Feldman, David, 1950—
How does aspirin find a headache?: an imponderables book / David Feldman.—1st ed.
p. cm
ISBN 0-06-016923-0
1. Curiosities and wonders. 2. Questions and answers. I. Title.
AG243.F42 1993
031.02—dc20 92-56221

ISBN 0-06-074094-9 (pbk.)

08 09 ❖/RRD 10 9 8 7 6 5

For Michele Gallery

Contents

Preface

We break down Imponderables into two categories. Type A Imponderables are the mysteries of everyday life that have always driven us nuts. Like why do lion tamers use kitchen chairs to fend off wild animals? Or why don't disc jockeys identify the titles and artists of the songs they play anymore?

Type B Imponderables are just as perplexing. These are mysteries we had never contemplated until our readers brought them to our attention. Believe us, we had never worried before about why the Muppets are left-handed, or why the Three Musketeers didn't carry muskets. But once we heard the questions, we *had* to find out the answers.

The readers of our last six *Imponderables* books have supplied us with scads of great Type A and Type B Imponderables. If we are stumped by a particularly challenging Imponderable, we convert it into a Frustable (short for Frustrating Imponderables) and ask readers to help bail us out. And in the Letters section, we let you enumerate our multifarious shortcomings.

We implore you to help us in our quest to stamp out all types of Imponderability, wherever they should lurk. If you are the first to submit an Imponderable we use in a book or provide the best solution to a Frustable, we offer you a grateful acknowledgment and a complimentary autographed copy of that book.

Want to contribute to this noble pursuit? See the last page to find out how you can join in. Don't be afraid. We're not carrying kitchen chairs.

Imponderables

Roaches In Your Car?!
Park It in a Kill 'Em ROACH GARAGE!!
OH NO!
Kill 'Em
NEW
Choose from CAPE COD or RAISED RANCH!
Roaches may hitch a ride...
...but it's a DEAD END street!

Why Don't We Ever See Cockroaches in Our Usually Crumb-Filled Cars?

Our correspondent, Manny Costa, wonders why an automobile, laden with assorted crumbs, wouldn't be a buffet paradise for our little scampering friends. Mary H. Ross, professor of entomology at Virginia Polytechnic Institute, isn't willing to state unequivocally that cockroaches are never found in cars, but she agrees it is rare. And she offers two main reasons why.

For one, cars may get too cold. Cockroaches dislike the cold and would refuse to stay in the car. Secondly, water is essential for cockroaches' survival and reproduction.

Richard Kramer, director of research, education, and technical resources at the National Pest Control Association, told us that while cockroaches require food every seven to ten days, they must take in water every three days. Perhaps a cockroach might be attracted to a stretch limousine with a leaky wet bar, but most of us don't drive limos.

If you really want to entice cockroaches into your automobile,

Kramer suggests scattering your empty beverage cans alongside your array of crumbs—you may be able to "support a cockroach infestation for a limited period of time."

Submitted by Manny Costa of Warwick, Rhode Island.

Why Does Barbie Have Realistic Nylon Hair While Ken Is Stuck with Plastic Hair or Painted Hair?

Poor Mattel is being attacked from all sides. Many feminists have criticized Barbie for setting up unrealistic expectations among girls about what their bodies should look like. Mattel answers, understandably, that Barbie was created to be a fashion doll, a model-mannequin suitable for hanging a variety of clothes upon. Of course, girls fantasize about themselves *as* Barbie, and this identification with the doll is precisely what the critics are worried about.

As if these complaints weren't enough for Mattel to worry about, here come six female *Imponderables* readers accusing the company of reverse discrimination. "What's the deal with Ken's hair?" they all wondered.

Informal chats with a gaggle of Barbie enthusiasts, both young and middle-aged, yielded the information that most girls are indifferent to Ken. To these fans, "Barbie doll" connotes visions of loveliness, while "Ken doll" evokes the image of the sterile figure atop wedding cakes.

Mattel's research indicates that there isn't much demand among girls for more realistic hair for Ken. Lisa McKendall, manager of marketing communications for Mattel, provides an explanation:

> In general, the most popular play pattern with fashion dolls among young girls is styling the hair. That is why long, combable hair is such an important feature of fashion dolls. Since the Ken doll's hair is short, there is much less to style and play with, so having "realistic" hair has not been as important.

Needless to say, "Ken hair" is much cheaper for Mattel to produce, particularly because painted hair doesn't have to be "rooted" to the top of the doll's head.

The choice of hairstyles for the Barbie lines is not taken casually. Meryl Friedman, vice-president of marketing for Barbie consumer products, told *Imponderables* that the length and texture of dolls' hair depends upon which "segment," or line of Barbies, Mattel is conceptualizing. Friedman reports that the best-selling doll in the history of Mattel is the "Totally Hair Barbie" line. The Barbie in the Totally Hair line is ten and one-half inches long—and the doll is only eleven and one half inches tall. In this particular segment, even Ken has combable, if short hair, as McKendall explains:

> . . . the Ken doll *does* have realistic-looking hair and actually comes with styling gel to create many different looks. A special fiber for the hair called Kankelon is produced specifically for us in Japan.

Friedman reports that in 1994, a Ken will be produced with longer hair.

Who says the men's liberation movement hasn't achieved anything?

Submitted by Dona Gray of Whiting, Indiana. Thanks also to Laura and Jenny Dunklee of Sutter Creek, California; Jessica Barmann of Kansas City, Missouri; Rebecca Capowski of Great Falls, Montana; and Nicole McKinley of Rochelle, Illinois.

On the U.S. Penny, Why Is the "o" in the "UNITED STATES oF AMERICA" on the Reverse Side in Lower Case?

Believe it or not, that little "o" is an artistic statement. According to Brenda F. Gatling, chief, executive secretariat of the United States Mint, the designer of the reverse side of the one-cent piece, Frank Gasparra, simply preferred the look of the little "o" alongside the big "F." And this eccentricity is not an anomaly; the Franklin half-dollar and several commemoratives contain the same, puny "o."

Submitted by Jennifer Godwin of Tyrone, Georgia.

What Do All the Chime Signals on Airlines Mean? Are They Uniform from Airline to Airline?

We might not be white-knuckle fliers anymore, but let's put it this way: We're closer to a pale pink than a full-bodied red. So we're not too happy when we find ourselves sitting next to fearful fliers. Why is our fate in life always to be seated alongside a middle-aged passenger taking his or her first flight? Invariably, our rowmates quake when they hear the landing gear go up. And more than one has reacted to the chime signals as if they were a death knell; one skittish woman knocked our Diet Coke off our tray when she heard the chimes. She assumed that the three-chime signal must signify that our flight was doomed. Actually, all that happened of consequence was that our pristine white shirt soon resembled the coat of a dalmatian.

But we always have been curious about the meaning of these chime codes, so we contacted the three largest airlines in the United States—American, United, and Delta—to ask if they would decode the mystery. We were surprised at how forthcoming they were. Nevertheless, for the first time in the history of *Imponderables*, we are going to withhold some of the information our sources willingly provided, for two reasons. First, airline chime-signals vary not only from airline to airline but from plane to plane within companies, and today's signals are subject to change in the future. Second, every airline *does* have a code to signify a true emergency, and the airlines aren't particularly excited about the idea of passengers decoding such a signal before the cockpit crew has communicated with flight attendants. Airlines are justifiably concerned about readers confusing emergency signals with innocuous ones and confusing one company's codes with another's. We agree.

Michael Lauria, an experienced pilot at United Airlines, told *Imponderables* that he has never had to activate an emergency chime signal. He is much more likely to sound one chime, to indicate that the cockpit wishes to speak to the first-class cabin attendant or (two chimes) to the coach flight attendants. Even if Lauria's passengers are enduring particularly nasty turbulence, chances are that the cry for help from the cockpit, expressed by the chimes, is more likely to be for a coffee or a soda than for draconian safety measures.

The number of chimes is not the only way of differentiating signals. Some United planes emit different tone frequencies: a lower-tone chime is heard for a passenger call than for a crew call, and a "bing bong" indicates a call from one flight attendant to another.

American Airlines uses different chime configurations to inform attendants when they should prepare for landing, remain seated with seat belts fastened, and call the cockpit crew. Although American does have a designated emergency signal, like other airlines' it is rarely used.

Delta Airlines features an array of different chime signals, which specify events during a flight. For example, when the "fasten seat belt" signs are turned off, a double high-low chime marks the event. These chimes also tell the flight attendants what elevation the plane has attained. Even during uneventful flights, there are periods of "sterile cockpits," when attendants are not supposed to disturb the cockpit crew except in an emergency. Sterile cockpits occur during takeoff and landing, and even though domestic airlines no longer allow smoking anywhere on the plane, some airlines still use the turning off "no smoking" sign as the marker for when the pilots can be contacted freely.

On most Delta planes, each phone station has a select tone, so that on a widebody plane, the flight attendant can recognize who is calling, and the flight crew can call any one or all of the flight attendant stations at one time. Alison Johnson, manager of aircraft interiors for Delta, told *Imponderables* that during an emergency, it is important for the flight crew to be able to speak to flight attendants without causing panic among passengers. Obviously, if the entire staff is briefed, a game plan can be established before informing passengers about a potential problem.

Submitted by Gabe Wiener of New York, New York. Thanks also to Dr. Richard Presnell of Augusta, Georgia.

Are Lions Really Afraid of Kitchen Chairs?

Give us a bazooka, a ten-foot pole, forty bodyguards, and excellent life insurance and hospitalization policies, and we might consider going into the ring with a lion.

Come to think of it, we think we'll still pass on it. But how in the h*@# did professional animal trainers choose such inappropriate tools as a whip and a kitchen chair? Why would a kitchen chair tame a lion? It doesn't even scare us!

At one time, animal trainers did use more forceful weapons against big cats. In his book *Here Comes the Circus,* author Peter Verney reports that the foremost trainer of the 1830s and 1840s, Isaac Van Amburgh, used heavy iron bars. Other trainers employed red-hot irons, goads, and even water hoses to control unmanageable beasts.

As far as we could ascertain, the considerably calmer instrument of the kitchen chair was introduced by the most famous lion tamer of the twentieth century, Clyde Beatty, who trained lions from 1920 until the late 1960s (when, ironically, he died of a car accident). His successor at the Clyde Beatty Circus, David Hoover, has strong feelings about the psychology of lions. Hoover believes that each lion has a totally different set of fears and motivations. For example, one lion he trained had a perverse fear of bass horns, while another went crazy when the circus's peanut roaster was operating.

Hoover believes that the only way for a human to control a lion is to gain psychological dominance over the animal—what he calls a "mental bluff." He also feels it important that the lion believe it couldn't harm the trainer. When he sustained injuries in the ring, Hoover always finished the act

> because the animal is operating under the assumption that he can't hurt you. If you leave the cage after the animal has injured you, then the animal knows he's injured you. You can't handle that animal anymore.

Hoover favored a blank cartridge gun over a whip. The purpose of the blanks was simply to disrupt the animal's concentration:

> They have a one-track mind. A blank cartridge goes off, and if you holler a command that the animal is familiar with, the animal will execute the command because he loses his original train of thought.

But what about the chair?

> The chair works the same way. The chair has four points of interest (the four legs). The animal is charging with the idea to tear the trainer apart. You put the chair up in his face. [When he sees the four legs of the chair] he loses his chain of thought, and he takes his wrath out on the chair and forgets he's after the trainer.

Ron Whitfield, lion tamer at Marine World Africa USA, told *Imponderables* that the chair is used more for theatrical reasons than for defense. If the instrument of the lion tamer is used as a tool of distraction rather than aggression, it makes sense to use flimsy props. Lions can be trained to bounce, swat at the chair—even to knock the chair out of the hands of the trainer. Ron assured us that if the lion wanted to attack, the chair would not offer any real protection.

Whitfield, who has trained lions for twenty-two years, has never used a chair. (He uses a stick and a crop whip.) The whip, he believes, is used as an extension of the hand of the trainer—to cue lions, who are lazy by nature. If they are sitting idly when they are supposed to be performing, a snap near them or a touch on their behinds will provide "motivation" to perform. And the whip provides negative reinforcement. Like a child, the lion learns that certain behaviors will induce a sting on the behind and will alter its behavior accordingly.

Even if lions have been performing in acts for years, they are still wild animals. Gary Priest, animal behavior specialist at the San Diego Zoo, reminds us that much of the behavior of even a "tamed" lion is instinctual and automatic. Like Hoover, Priest emphasizes the necessity for trainers to demonstrate a lack of fear of the animal. Without intervention of some kind, lions would revert to the appetitive cycle (crouching, eyes squinting, ears pinned back, lowering into a crouch, and springing) characteristic of lions on the hunt.

If trainers run or show fear, Priest explains, lions will think of them as prey. But if you approach the lion before it snares you, the appetitive cycle is disrupted. In the wild, no prey *would* approach a lion, so the cat is not genetically encoded to respond to this aggressiveness. (Indeed, Priest told us that if you ever encounter a lion in the wild, especially one that is starting to crouch, do *not* run away. Instead, run *toward* the lion, yelling "bugga bugga bugga," or some such profundity. Says Priest: "This will probably save your life," as the lion usually retreats when confronted.)

Priest thinks that the whip and chair are a good combination, with the noise from the whip a particularly good distraction and the chair allowing the trainer to approach the lion and still have some distance and (minimal) protection.

The one part of this Imponderable we would have loved to unravel is how and why Clyde Beatty thought of the idea of using a kitchen chair in the first place. Did he have a scare with a cat one day when the chair was the only object handy? Was he sitting on a chair when a lion attacked? Or did he just want an excuse to drag his favorite kitchen chair around the world with him?

Submitted by Steven Sorrentino of West Long Branch, New Jersey.

Why Is Pistachio Ice Cream Colored Green?

We first were asked this question in 1988 but decided not to answer it because we thought the answer was obvious: Pistachio nuts *are* green (sort of). Sure, you have to liberate the nutmeats from their ivory or red-colored shells (please don't ask us why pistachios are dyed red—see *Imponderables* for the thrilling answer), and then detach the thin, reddish-brown husk. Underneath its wrappers, though, is a yellowish nut with an obvious green tinge.

But this question seems to be on the minds of North Americans everywhere, and we are the last to deprive our readers of the knowledge that can set them free. We first contacted Ed Marks, an ice cream expert and historian, who told us that the first reference he could find to green pistachio ice cream was in *The Standard Manual for Soda and Other Beverages* (1897), by A. Emil Hiss. The directions state: "Color green with a suitable delicate color."

Why was green chosen? Marks has his ideas:

> The use of color in ice cream and other food items is predicated on two things: to make the food appealing to the eye and to generally make a processed item appear more closely to its natural state. One would presume that the use of green serves either or both of these purposes. I always associated green with the color of the pistachio nutmeat, although the truth is that the term "yellow-green" is a more apt description. Perhaps the early ice cream makers made a subjective decision that green was more appropriate than yellow.

Donald Buckley, executive director of the National Ice Cream Retailers Association, echoed Marks's sentiments and added that the color green had relatively little competition in early ice cream fountains. Yellow was already "taken" by vanilla. The average person probably associates the color green most closely with mint, not then popular as an ice cream flavor. (Now, green is a popular color in mint chocolate chip ice cream.)

Kathie Bellamy, of Baskin-Robbins, indicated that consumers are very conscious of whether the color of an ice cream simulates the "public perception of pistachio." Note that most commercial pistachio ice creams, including Baskin-Robbins', are invariably a pale green, and that considerable research is conducted on "color appeal." After all, with inexpensive food colorings, pistachio could just as easily be colored chartreuse, if the public would buy it.

Submitted by Lynda Frank of Omaha, Nebraska. Thanks also to Morgan Little of Austin, Texas; Bob Muenchow of Meriden, Connecticut; and many others.

Why Are Graves Six Feet Deep and Who Determined They Should Be That Deep?

Graves haven't always been that deep. Richard Santore, executive director of the Associated Funeral Directors International, told *Imponderables* that during the time of the Black Plague in Europe, bodies were not buried properly or as deep as they are today. These slovenly practices resulted in rather unpleasant side effects. As soil around the bodies eroded, body parts became exposed, which explains the origins of the slang term "bone yard" for a cemetery. Beside the grossness content, decomposing flesh on the surface of the earth did nothing to help the continent's health problems.

England, according to Santore, was the first to mandate the six-foot-under rule, with the idea that husband and wife could be buried atop each other, leaving a safe cushion of two feet of soil above the buried body, "the assumption being that if each casket was two feet high, you would allow two feet for the husband, two feet for the wife, and two feet of soil above the last burial." At last, there were no bones in the bone yard to be found.

The six-foot rule also puts coffins out of reach of most predators and the frost line. Of course, caskets could be buried even deeper, but that would be unlikely to be popular with gravediggers, as Dan Flory, president of the Cincinnati College of Mortuary Science, explains:

> Six feet is a reasonable distance for the gravedigger [to shovel] and is usually not deep enough to get into serious water or rock trouble.

Submitted by Patricia Arnold of Sun Lakes, Arizona. Thanks also to Deone Pearcy of Tehachapi, California.

Why Doesn't Ham Change Color When Cooked, Like Other Meats?

Let's answer your Imponderable with a question. Why isn't ham the same color as a pork chop, a rather pallid gray?

The answer, of course, is that ham is cured and sometimes smoked. The curing (and the smoking, when used) changes the color of the meat. You don't cook a ham; you reheat it. According to Anne Tantum, of the American Association of Meat Processors, without curing, ham would look much like a pork chop, with perhaps a slightly pinker hue.

Curing was used originally to preserve meat before the days of refrigerators and freezers. The earliest curing was probably done with only salt. But salt-curing alone yields a dry, hard product, with an excessively salty taste.

Today, several other ingredients are added in the curing process, with two being significant. Sugar or other sweeteners are added primarily for flavor but also to retain some of the moisture of the meat that salt would otherwise absorb. Sugar also plays a minor role in fighting bacteria.

For our purposes, the more important second ingredient is nitrites and/or nitrates. Sodium nitrate is commonly injected into the ham, where it turns into nitrite. Nitrite is important in fighting botulism and other microorganisms that spoil meat or render it rancid. Nitrites also lend the dominant taste we associate with cured meat (bacon wouldn't taste like bacon without nitrites).

Unfortunately, for all the good nitrite does in keeping ham and other meats from spoiling, a controversy has arisen about its possible dark side. When nitrites break down, nitrous acid forms. Combined with secondary amines (an ammonia derivative combining hydrogen and carbon atoms), nitrous acid creates nitrosamines, known carcinogens. The debate about whether nitrosamines develop normally during the curing process is still swirling.

But nitrite was used to cure pork even before these health benefits and dangers were known, because it has always been valued as an effective way to color the meat. Nitrites stabilize the color of the muscle tissues that contain the pink pigment we associate with ham, as do some of the other salts (sodium erythorbate and/or sodium ascorbate) that help hasten the curing process.

Most curing today is done by a machine, which automatically injects a pickle cure of (in descending order of weight) water, salt, sweetener, phosphate, sodium erythorbate, sodium nitrate, and sodium nitrite. Usually, multiple needles are stuck in the ham; the more sophisticated machines can inject even bone-in hams. After this injection hams are placed in a cover pickle, where they sit for anywhere between a few days and a week. Hams that sit in cover pickle sport rosier hues than those that are sent directly to be cooked.

Hams are smoked at very low temperatures, under 200 degrees Fahrenheit, usually for five to six hours. Some cooked hams ("boiled ham" is a misnomer, as few hams are ever placed in water hotter than 170 degrees Fahrenheit) are cooked unsmoked in tanks of water and tend to be duller in color. These hams are usually sold as sandwich meat.

One of the reasons why hams are beloved by amateur cooks is that, like (cured) hot dogs, they are near impossible to undercook or overcook. More than a few Thanksgiving turkeys have turned into turkey jerky because cooks didn't know when to take the bird out of the oven. Luckily, hams are precooked for us. We might have to pay for the privilege, but it is hard for even noncooks to ruin their texture or tarnish their pinkish color.

Submitted by Dena Conn of Chicago, Illinois.

Why Does Warm Milk Serve as an Effective Sleep-Inducer for Many People?

Scientists haven't been able to verify it, but there is some evidence to support the idea that milk actually might induce sleep. Milk contains tryptophan, an amino acid, which is the precursor of a brain transmitter, serotonin, which we know has sedative qualities.

Recently, L-tryptophan supplements, which had gained popularity as a sleeping aid, were banned by the Food and Drug Administration because they caused severe reactions in some users, including eosinophilia, an increase in the number of white blood cells. Earlier research confirmed that L-tryptophan did help many people get to sleep faster than a placebo.

But does cow's milk contain enough tryptophan to induce sleepiness? This has yet to be proven. Representatives of the dairy industry, who might be the first to claim such a benefit for milk, are reluctant to do so and are openly skeptical about its sleep-inducing qualities. Jean Naras, a media relations specialist at the American Dairy Association, although dubious about the sedative effects of milk, cited research that indicated it might take a dose as high as a half-gallon to provide any sleep benefits. And with this quantity of milk intake, your bladder might argue with your brain about whether you really want to sleep through the night.

And does warm milk promote sleep any better than cold milk? It does if you believe it does, but none of the experts we consulted could provide a single logical reason why it should.

Submitted by R.W. Stanley of Bossier City, Louisiana.

Why Doesn't Glue Get Stuck in the Bottle?

There are two basic reasons:

1. In order for glue to set and solidify, it must dry out. Latex and water-based glues harden by losing water, either by absorption into a porous substrate (the surface to be bonded) or by evaporation into the air. The glue bottle, at least if it is capped tightly, seals in moisture.

2. Different glues are formulated to adhere to particular substrates. If the glue does not have a chemical adhesion to the substrate, it will not stick. For example, John Anderson, technical manager for Elmer's Laboratory (makers of Elmer's Glue-All), told us that the Elmer's bottle, made of polyethylene, does not provide a good chemical adhesion for the glue.

Even when the cap is left off, and the glue does lose water, the adhesion is still spotty. We can see this effect with the cap of many glue bottles. In most cases, dried glue can and does cake onto the tip after repeated uses. But Anderson points out that the adhesion is "tenuous," and one can easily clean the top while still wet and remove the glue completely. Likewise, if you poured Elmer's on a drinking glass, it might adhere a little, but you could easily wipe it off with a cloth or paper towel, because the glue cannot easily penetrate the "gluee."

Submitted by Jeff Openden of Northridge, California.

Early efforts to get subjects to smile~
WHAT?!
BLEAH!
OK ~ NOW SAY "LIMBURGER CHEESE"!
OH MY!
YUCK
ICK!
-weren't too successful!

Why Don't People in Old Photographs Ever Seem to Smile?

Sometimes, the more you delve into an Imponderable, the murkier it becomes. We asked about twenty experts in photography and photographic history, and the early responses were fairly consistent: The subjects in old photographs weren't all depressed; the slowness of the exposure time was the culprit. In some cases, the exposure time in early daguerreotypes was up to ten minutes. Typical was the answer of Frank Calandra, secretary/treasurer of the Photographic Historical Society:

> Nineteenth-century photographic materials were nowhere near as light-sensitive as today's films. This meant that instead of the fractional second exposure times we take for granted, the pioneer photographers needed several minutes to properly set an image on a sensitized plate. While this was fine for landscapes, buildings and other still-lifes, portraits called for many tricks to help subjects hold perfectly still while the shutter was open. (The first cameras had no shutter. A cap was placed over the lens and the photographer would remove it to begin the exposure and replace it when time was up.)

> Holding a smile for that length of time can be uncomfortable; that's why you see the same somber look on early portraits. That's what a relaxed face looks like.

It that's so, Frank, we'll look jittery, anytime.

Of course, the problem with trying to hold a smile for a long period of time isn't only that it is difficult. The problem is that the smile looks phony. Photographer Wilton Wong told *Imponderables* that even today,

> A good portraitist will *not* ask subjects to smile and have them hold it even for more than a few seconds, as the smile starts looking forced. With the long exposures of old, the smiles would look phony and detract from the photo. Look at yourself in the mirror with a thirty-second smile on your face!

The stationary of the Photographic Historical Association depicts a head clamp, which, although it looks like an instrument of torture, was used during the early days of photography to prevent a subject's head from moving while being photographed. In order to avoid blurring, subjects were forced to fix their gaze during the entire session. Iron braces were also utilized to keep the neck and trunks of subjects from moving. According to photographer Dennis Stacey,

> Sometimes all three brace methods were used, and in the case of young children, a sash was employed to tie them to the fixture where they sat to assist in holding them motionless.

In his book *The History of Photography from 1839 to the Present,* Beaumont Newhall recounts many anecdotes about the hardships caused by long exposure times. American inventor Samuel Morse, who was sent an early prototype by Daguerre (the French artist who pioneered photography), sat his wife and daughter down "from ten to twenty minutes" for each photograph. Newhall also describes the travails of an anonymous "victim," who suffered through an excruciating single shot:

> . . . he sat for eight minutes, with the strong sunlight shining on his face and tears trickling down his cheeks while the operator promenaded the room with watch in hand, calling out the time every five seconds, till the fountains of his eyes were dry.

We were satisfied with this technological explanation for unsmiling subjects until we heard from some dissenters. Perhaps the most vehement is Grant Romer, director of education at the George Eastman House's International Museum of Photography. Romer told us that the details of the daguerreotype process were announced to the public on August 19, 1839, and that only immediately after this announcement were exposures this long. By 1845, exposure time was down to six seconds. Yes, Romer admits, often photographers did utilize longer exposure times, but the technology was already in place to dramatically shorten the statuelike posing of subjects.

So could we find alternative explanations for the moroseness of early photographic subjects? We sure could. Here are some of the more plausible theories:

1. *Photographs were once serious business.* Joe Struble, assistant archivist at the George Eastman House, told us that the opportunity to have a photographic portrait was thought of as a once-in-a-lifetime opportunity. And it isn't as if the Victorian era was one where goofiness was prized. Roy McJunkin, curator of the California Museum of Photography at the University of California, Berkeley, feels that the serious expressions embody a "Victorian notion of dignity—a cultural inclination to be seen as a serious, hardworking individual." Photographer John Cahill told us that even in the 1990s, it is not unusual for a European or Middle Eastern subject to thank him for taking even a casual snapshot.

2. *Early subjects were imitating the subjects of portrait painters.* Daguerre was himself a painter, and early photographers saw themselves as fine artists. Jim Schreier, a military historian, notes that subjects in this era did not smile in paintings. John Hunisak, professor of art history at Middlebury College, who concurs with Schreier, adds that early photographs of landscapes also tried to mimic still-life paintings.

3. *"Technology and social history are intimately interwoven."* This comment, by Roy McJunkin, indicates his strong feeling that George Eastman's invention of roll film, and the candid camera (which could be carried under a shirt or in a purse), both before the end of the nineteenth century, eventually forced the dour expressions of early photographic subjects to turn into smiles. The conventions of the early portrait pictures were changed forever once families owned their own cameras.

In the 1850s, according to McJunkin, the average person might have sat for a photographic portrait a few times in his or her life. With roll film, it became possible for someone oblivious to the techniques of photography to shoot pictures in informal settings and without great expense.

4. *Early photographs were consciously intended for posterity.* When asked why people didn't smile in old photographs, Grant Romer responded, "Because they didn't want to." Romer was not being facetious. Photographic sessions were "serious business" not only because of their rarity and expense but because the photographs were meant to create documents to record oneself for posterity. Rather, he emphasizes that until the invention of the candid camera, photographers might have asked subjects to assume a pleasant expression, but the baring of teeth or grinning was not considered the proper way to record one's countenance for future generations.

5. *Who wants to bare bad teeth?* Romer does not discount the poor dental condition of the citizenry as a solid reason to keep the mouth closed. As Tampa, Florida, photographer Kevin Newsome put it:

> Baking soda was the toothpaste of the elite. Just imagine what the middle class used, if anything at all.

6. *Historical and psychological explanations.* As compelling as all of these theories are, we still feel there are psychological, historical, and sociological implications to the expressions of the subjects in old photographs. In "The Photography of History," a fascinating article in *After Image,* Michael Lesy discusses the severe economic depression that began in 1836 and lasted six years. Photography was brought to the United States in the thick of it. Lesy observes that early American photographers were not seen as craftsmen or artists but as mesmerists and phrenologists (belief in both was rampant):

> The daguerreotypists were called "professor" and were believed to practice a character magic that trapped light and used the dark to reveal the truth of a soul that shone through a face. These men may have been opportunists, but they moved through a population that lived in the midst of a commercial, political, and spiritual crisis that lasted a generation that ended with carnage and assassination. The craft they practiced and the pictures they made

were the result not only of the conventional rationalism of an applied technology, but of irrational needs that must be understood psychologically.

In other words, this was not a period when photographers enticed subjects to yell "cheese," or put a devil sign over the heads of other subjects.

By the time George Eastman introduced the roll camera, Victorian morality was waning. People who bought hand cameras, for the most part, were not the generation that suffered through the privations of the Civil War. At the turn of the century, there was a new middle class, eager to buy "cutting edge" technology.

As a benchmark, Roy McJunkin asks us to look at photographs of Queen Victoria and President McKinley and compare them with the visage of Teddy Roosevelt, a deft politician who consciously smiled for the camera: "He was the first media president—he understood what a photo opportunity was and took advantage of it." The politics of joy was born, and with it a new conviction that even the average working stiff had a right not specified in the Constitution—not only to the pursuit of happiness but to happiness itself.

We have one confession to make. Our research has indicated that smiles in old photographs, while uncommon, did exist. Dennis Stacey told us that more than a few existing Victorian photographs show the sitter "grinning or smiling, usually with the mouth closed." Grant Romer has an 1854 daguerreotype at the Eastman House of a woman standing on her head in a chair, smiling.

Romer admits that she was clearly defying the conventions of the period. But then, a little smile made Mona Lisa daring in her day, too.

Submitted by Ken Shafer of Traverse City, Michigan. Thanks also to Donna Yavelak of Norcross, Georgia; Rick Kaufhold of Dayton, Ohio; Sally Esposito of Las Vegas, Nevada; and Bart Hoss of Kansas City, Missouri.

Why Did Men Thrust Their Right Hand into Their Jackets in Old Photographs?

Most of the photo historians we contacted discounted what we considered to be the most likely answer: These subjects were merely imitating Napoleon and what came to be known as the Napoleonic pose. Maggie Kannan, of the department of photographs at the Metropolitan Museum of Art, and other experts we contacted felt that many of the reasons mentioned in the Imponderable above were more likely.

Just as subjects couldn't easily maintain a sincere smile during long exposure times, so was trying to keep their hands still a challenge. Frank Calandra wrote us:

> The hand was placed in the jacket or a pocket or resting on a fixed object so that the subject wouldn't move it [or his other hand] and cause a blurred image. Try holding your hands at your sides motionless for fifteen minutes or so—it's not easy.

Grant Romer adds that this gesture not only solved the problem of blurring and what to do with the subject's hands while striking a pose but forced the subject to hold his body in a more elegant manner.

Still, if these technical concerns were the only problem, why not thrust both hands into the jacket? Or pose the hands in front of the subject, with fingers intertwined? John Husinak assured us that this particular piece of body language was part of a trend that was bigger and more wide-ranging than simply an imitation of Napoleon.

Early portrait photographers understood the significance of particular gestures to the point where they were codified in many journals and manuals about photography. Some specific examples are cited in an article by William E. Parker in *After Image,* an analysis of the work of early photographer Everett A. Scholfield. Parker cites some specific examples: Two men shaking hands or touching each other's shoulders "connoted familial relationship or particular comradeship"; if a subject's head was tilted up with the eyes open or down with eyes closed, the photographer meant "to suggest speculative or contemplative moods."

Harry Amdur, of the American Photographic Historical Society, told us that early photographers tried to be "painterly" because they wanted to gain respect as fine artists. Any survey of the portrait paintings of the early and mid-nineteenth century indicates that the "hand-in-jacket" pose was a common one for many prominent men besides Napoleon.

Another boon to the Napoleonic pose was the invention of the *carte de visite,* a photographic calling card. Developed in France in the 1850s, small portraits were mounted on a card about the size of today's business card. Royalty and many affluent commoners had their visages immortalized on *cartes.* In France, prominent figures actually sold their *cartes*—ordinary citizens collected what became the baseball cards of their era. *Cartes de visite* invaded the United States within years.

These photos were far from candid shots. Indeed, Roy McJunkin told *Imponderables* that *carte de visite* studios in the United States used theatrical sets, and that subjects invariably dressed in their Sunday best. The hand-in-jacket pose was only one of many staged poses, including holding a letter or bible, holding a gun as if the subject were shooting, or pointing to an unseen (and usually nonexistent) point or object.

Some of the pretensions of this period were downright silly—silly enough to inspire Lewis Carroll to write a parody of the whole enterprise. In Carroll's poem, actually a parody of Longfellow's *Hiawatha,* Hiawatha is transformed into a harried, frustrated portrait photographer:

> From his shoulder Hiawatha
> Took the camera of rosewood,
> Made of sliding, folding rosewood;
> Neatly put it all together,
> In its case it lay compactly,
> Folded into nearly nothing;
> But he opened out the hinges,
> Pushed and pulled the joints and hinges,
> Till it looked all squares and oblongs,
> Like a complicated figure
> In the second book of Euclid,
>   This he perched upon a tri-pod—
>   Crouched beneath its dusty cover—

Stretched his hand enforcing silence—
Mystic, awful was the process,
  All the family in order
Sat before him for their pictures:
Each in turn, as he was taken,
Volunteered his own suggestions.
  First the governor, the father:
He suggested velvet curtains
Looped about a messy Pillar;
And the corner of a table.
He would hold a scroll of something
Hold it firmly in his left hand;
He would keep his right hand buried
(Like Napoleon) in his waistcoat;
He would contemplate the distance
With a look of pensive meaning,
As of ducks that die in tempests.
Grand, heroic was the notion:
Yet, the picture failed entirely:
Failed, because he moved a little,
Moved because he couldn't help it!

Who would have ever thought of Lewis Carroll summarizing the answers to an Imponderable, while simultaneously contemplating the plight of a Sears portrait photographer?

Submitted by Donald McGurk of West Springfield, Massachusetts. Thanks also to Wendy Gessel of Hudson, Ohio, and Geoff Rizzie of Cypress, California.

Why Are Carpenter's Pencils Square?

Two reasons. Carl Reichenbach, product manager at pencil giant Dixon-Ticonderoga, told *Imponderables* that the square shape enables carpenters to draw thin or thick lines more easily than with conventional pencils.

But a more pressing point: If we drop a pencil from our desk, it's not a big deal to lean over and pick it up from the floor. However, what if we happen to drop a pencil from a beam on the thirty-fourth floor of a construction site? Or the roof of a home? As Ellen B. Carson of Empire Berol USA put it, "The carpenters' pencils are produced in a square shape so they won't roll off building materials."

Submitted by Nate Woodward of Seattle, Washington.

Why Don't Windshield Wipers in Buses Work in Tandem Like Auto Wipers?

Hearing that two *Imponderables* readers were obsessed with this question made us feel less lonely. We've always wondered whether we were the only ones bugged by the infernal racket and displeasing look of two huge, awkward, asymmetrical windshield wipers churning away on rainy trips.

The answer turns out to be simple, if technical. Most automobiles use one motor to power two windshield wipers. With bigger windshields and blades, the two bus wipers are driven by separate, independent motors, so the movement of the two blades is not coordinated.

Isn't there any way to get the two wipers to work together? Sure, for a cost, as Karen Finkel, executive director of the National School Transportation Association, explains:

> A larger motor to accommodate the larger blade and windshield could be developed. However, there isn't a reason to synchronize the wipers so it hasn't been done.

Hmmm. Not driving us nuts, we guess, isn't reason enough to change the status quo.

Submitted by P.M. Cook of Lake Stephens, Washington. Thanks also to Karyn Heckman of Greenville, Pennsylvania.

Why Were Athos, Porthos, and Aramis Called the Three Musketeers When They Fought with Swords Rather Than Muskets?

The Three Swordsmen sounds like a decent enough title for a book, if not an inspiring name for a candy bar, so why did Dumas choose *The Three Musketeers*? Dumas based his novel on *Memoirs of Monsieur D'Artagnan,* a fictionalized account of "Captain-Lieutenant of the First Company of the King's Musketeers." Yes, there really was a company of musketeers in France in the seventeenth century.

Formed in 1622, the company's main function was to serve as bodyguard for the King (Louis XIII) during peacetime. During wars, the musketeers were dispatched to fight in the infantry or cavalry; but at the palace, they were the *corps d'élite.* Although they were young (mostly seventeen to twenty years of age), all had prior experience in the military and were of aristocratic ancestry.

According to Dumas translator Lord Sudley, when the musketeers were formed, they "had just been armed with the new flintlock, muzzle-loading muskets," a precursor to modern rifles. Unfortunately, the musket, although powerful enough to pierce any armor of its day, was also extremely cumbersome. As long as eight feet, and the weight of two bowling balls, they were too unwieldy to be carried by horsemen. The musket was so awkward that it could not be shot accurately while resting on the shoulder, so musketeers used a fork rest to steady the weapon. Eventually, the "musketeers" were rendered musketless and relied on newfangled pistols and trusty old swords.

Just think of how muskets would have slowed down the derring-do of the three amigos. It's not easy, for example, to slash a sword-brandishing villain while dangling from a chandelier, if one has a musket on one's back.

Submitted by John Bigus of Orion, Illinois.

Why Don't Public Schools Teach First Aid and CPR Techniques?

Wouldn't our world be a safer place if every high school required students to take a class in life-saving techniques? Reader Charles Myers sure thought so, so we tried to find out why CPR isn't a part of schoolroom curriculums in most communities. Considering the violence in our schools, the argument needn't be made that such training would only be usable in the "outside world." In fact, this issue recently has become a hot topic among educators, particularly because of concerns about the response times of fire, paramedic, police, and other emergency medical services in many communities.

All of the health officials we contacted felt that CPR and first aid training in public schools could be valuable, but they provided a litany of reasons why we shouldn't expect to see it in the near future. Why not?

1. *Money.* Most school systems are riddled with financial problems. CPR training is labor-intensive. While a normal classroom might have a ratio of twenty-five or thirty students per teacher, CPR requires a six-to-one or eight-to-one ratio. And training teachers to learn and then teach CPR costs money.

2. *Liability problems.* "America," says Bill Powell, prehospital emergency training coordinator for Booth Memorial Medical Center in Flushing, New York, "is the land of the suing." What if a student botches a rescue operation? Would the school be legally and financially liable for poor training?

3. *No one should be forced to administer CPR.* Several of our sources indicated that although it is an admirable goal to have every student learn first aid techniques, in practice it might not be a good idea to force those who don't desire to learn or who are incapable of administering them properly. Georgeanne Del Canto, director of health services for the Brooklyn, New York, Board of Education, told *Imponderables* that good intentions notwithstanding, asking every student to learn first aid would be a little like asking every schoolkid to go out for the wrestling team: Too many students would be insufficiently strong, energetic, or limber to apply CPR adequately, and more than a few would be too squeamish.

If we required all students to learn CPR, we would be forcing them to learn a technique that they would not be obligated to use outside of school. In most states, Bill Powell told us, lay citizens are under no legal obligation to act as a Good Samaritan (in most states, physicians do have such an duty), even in life-threatening situations.

4. *Motivation of students.* Powell isn't too sanguine about the desire of students to learn CPR properly. Why should they pay any more attention to first aid training than they do to math or history? Most non-health professional students in CPR courses are people who are friends or family of someone who they fear is at risk; few learn CPR out of sheer altruism or an abstract academic interest.

5. *Time.* CPR certifications must be renewed every two years, not so much because first aid techniques change but because most students, thankfully, never get a chance to apply their lessons in real life. Constant retraining of students might be another financial and labor drain on schools, although this problem could be ameliorated by introducing the subject in the junior year of high school, thus saddling colleges with the task of recertification.

6. *Training of trainers.* "A little knowledge is a dangerous thing," emphasizes Ira Schwartz, project director of the New York State Regents Advisory Committee on Community Involvement. Schwartz and others we talked to thought that finding a ready supply of teachers qualified to

teach CPR, or training nonqualified teachers to do so, was a major hurdle.

7. *Competition.* CPR isn't the only health item clamoring to be included in public school curriculums. Although she joined in the chorus of educators who thought that universal CPR training would be a noble idea, Arlene Sheffield, director of the school health demonstration program for the New York State Education Department, told us that many educators believe that teaching students about bulimia, anorexia nervosa, and child abuse has a higher priority. Sheffield's program, for example, focuses on eliminating drug abuse among students, a subject of more pressing importance to students, educators, and parents than CPR. As Sheffield puts it,

> The pool of money and time available for schooling in any given subject is finite and CPR and first aid have a lot of healthy competition.

CPR training is not totally abandoned in our public schools. Many schools offer kids training on an elective basis; community groups and hospitals offer low-cost courses. And a few public school systems, such as Seattle's, find the time, money, and training resources to offer all students CPR courses.

Charles Myers was not alone in wondering if some of that time he (and, let us admit, we) wasted in school might have been better spent learning a skill that could save others' lives. When we asked Emmanuel M. Goldman, former publisher of *Curriculum Review,* this Imponderable, he echoed our sentiments exactly:

> As to why CPR isn't taught, at least in high school: It beats me. Probably because it is such a logical, desirable, and useful skill.

Submitted by Charles Myers of Ronkonkoma, New York.

Why Do Peanuts in the Shell Usually Grow in Pairs?

Botany 101. A peanut is not a nut but a legume, closer biologically to a pea or a bean than a walnut or pecan. Each ovary of the plant usually releases one seed per pod, and all normal shells contain more than one ovary.

But not all peanut shells contain two seeds. We are most familiar with Virginia peanuts, which usually contain two but occasionally sprout mutants that feature one, three, or four. Valencia and Spanish peanuts boast three to five seeds per shell.

Traditionally, breeders have chosen to develop two-seeded pods for a practical reason: Two-seeders are much easier to shell. According to Charles Simpson, of Texas A & M's Texas Agricultural Experiment Station, there is little taste difference among the varieties of peanuts, but the three-seed peanuts are quite difficult to shell, requiring tremendous pressure to open without damaging the legume. We do know that patrons of baseball games wouldn't abide the lack of immediate gratification. They'd much rather plop two peanuts than three into their mouths, at least if it means less toil and more beer consumption.

Submitted by Thad Seaver, A Company, 127 FSB.

Why Are Children Taught How to Print Before They Learn Cursive Handwriting?

While most of us were taught how to print in kindergarten and learned how to write in late second or third grade, this wasn't always the case. Until the early 1920s, children were taught only cursive handwriting in school. Margaret Wise imported the idea of starting kids with manuscript writing (or printing) from England in 1921, and her method has become nearly universal in North America ever since.

Wise used two arguments to promote the radical change: With limited motor skills, it was easier for small children to make print legible; and print, looking more like typeset letters than cursive writing, would enable children to learn to read faster and more easily. Subsequent experimental research has confirmed that Wise's suppositions were correct.

In the seventy years since Wise revolutionized kindergarten penmanship, other reasons for teaching children printing have been advanced: Print is easier for teachers and students to read; students learn print more quickly and easily than cursive writing; and despite protestations from some, children can print as fast as they can write. While adults tend to write faster than they can print, experiments have indicated that this is only true because most adults rarely print; those who print as a matter of course are just as fast as cursive writers.

We have pored over many academic discussions about children's writing and haven't found anyone strenuously objecting to teaching children how to print first. What surprised us, though, is the lack of reasoned justification for weaning children from printing and, just after they have mastered the technique, teaching them cursive writing. After reviewing the literature on the subject, Walter Koenke, in an article in *The Reading Teacher*, boiled the rationales down to two—tradition and parental pressure:

> Since printing can be produced as speedily as cursive handwriting while being as legible and since it is obvious that the adult world generally accepts printing, it seems that the tradition rather than research calls for the transition from some form of printing to cursive handwriting.

A litany of justifications for cursive writing has been advanced, but none of them holds up. If print is easier to read and write, why do we need to learn cursive script? Why do we need to teach children duplicate letter forms when there is hard evidence that the transitionary period temporarily retards students' reading and compositional ability? (One study indicated that for each semester's delay in introducing cursive handwriting, students' compositional skill improved.)

In a wonderful article, "Curse You, Cursive Writing," the University of Northern Iowa's Professor Sharon Arthur Moore argues passionately that there is no need to teach children cursive writing and rebuts most of the arguments that its proponents claim. It is not true, as conventional wisdom might have it, that cursive writing is harder to forge than manuscript print, nor is it true that only cursive writing can be a valid signature on legal documents (X can still mark the spot).

Moore feels that parental pressure and a belief that cursive is somehow more "grown up" or prestigious than print permeates our society and leads to an unnecessary emphasis on cursive style:

> From the time they enter school, children want to learn to "write"; near the end of second or the beginning of third grade, the wish comes true. The writing done to that point must not be very highly valued, or why would there be such a rush to learn to do "the real thing"? . . . perceptions are so much more powerful than reality at times that it may not even occur to people to question the value of cursive writing.

Why, she argues persuasively, do business and legal forms ask us to "please print carefully"? The answer, of course, is that even adults print more legibly than they write cursively. If cursive is superior, why aren't cursive typefaces for typewriters and computer printers more popular? Why aren't books published in cursive?

Moore, like us, can't understand the justification for teaching kids how to write, and then changing that method in two years for no pedagogical purpose. She endorses the notion of teaching cursive writing as an elective in eighth or ninth grade.

Several handwriting styles have been advanced to try to bridge the gap between manuscript and cursive styles—most prominently, the "D'Nealian Manuscript," which teaches children to slant letters from the very beginning and involves much less lifting of the pencil than standard printing. Proponents of the D'Nealian method claim that their style requires fewer jerky movements that may prove difficult and time-consuming for five- and six-year-olds and eases the transition from manuscript to cursive by teaching kids how to "slant" right away. And the D'Nealian method also cuts down the reversal of letters that typifies children's printing. It is far less likely that a child, using D'Nealian, will misspell "dad" as "bab," because the "b" and "d" look considerably different. In standard manuscript, the child is taught to create a "b" by making a straight vertical line and then drawing a circle next to it. But in D'Nealian, the pencil is never picked up: the straight but slanted vertical line is drawn, but the "circle" starts at the bottom of the line, and the pencil is brought around and up to form what they call the "tummy" of the "b."

We'll leave it to the theorists to debate whether the D'Nealian, or more obscure methods, are superior to the standard "circle stick" style of manuscript. But we wish we could have found a clearer reason why it's necessary to change from that style into cursive writing—ever. We couldn't argue the cause more eloquently than Janice-Carol Yasgur, an urban elementary schoolteacher:

> Just as kids begin to get competent in printing their thoughts, we come along and teach them cursive—and what a curse it is! Now they devote more of their energy to joining all the letters together than to thinking about what they're trying to communicate, so that it's a total loss: It's impossible to make out their scribbles; but even if you can, it's impossible to figure out what they're trying to say. It's a plot to keep elementary schoolteachers in a state of permanent distress.

Submitted by Erin Driedger of Osgoode, Ontario.

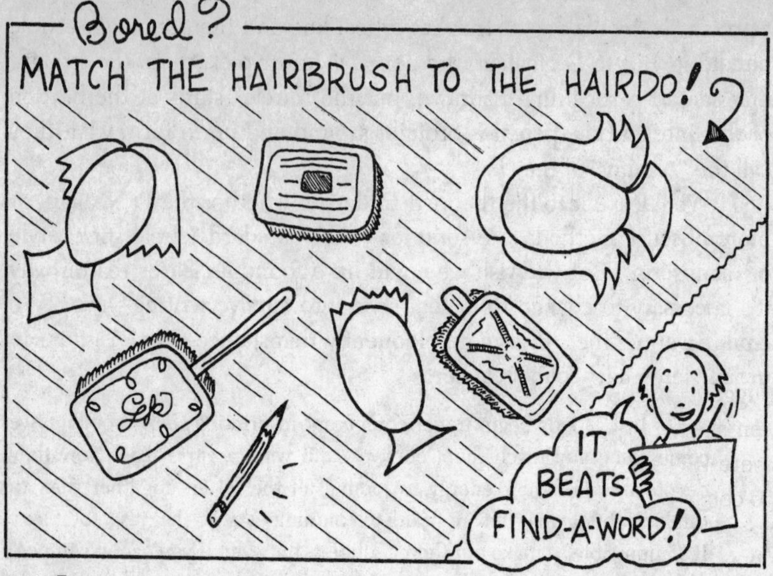

Why Do Most Women's Hairbrushes Have Long Handles When Men's Hairbrushes Have Short Handles or No Handles at All?

Why are men deprived of the graceful, long handles on women's hairbrushes? According to the experts we contacted, the answer seems to be that the longer the hair of the user is, the longer the handle of the brush should be. Carmen Miller, product manager of Vidal Sassoon brushes and combs division, explains:

> Traditionally, men have used what is referred to as a "Club" brush—a wide-based brush with densely packed bristles and a shorter length handle. This brush is best used for smoothing hair, not texturizing or detangling, as most women's brushes are used for. Since men usually have closely cropped hair, they need to use a brush closer to the scalp to effectively smooth their hair.

This response, of course, begs the question of whether Annie Lennox and Sinead O'Connor use long- or short-handled brushes. Or imagine the plight of Daniel Day Lewis, in *Last of the Mohicans*, ferreting the burrs out of his hair with a handleless brush. Fabio could use a long handle too.

Miller indicates that the shape of the man's hand, as well as the shortness of his hair, is a consideration in handle length:

> The [short] handle was designed to allow a man's hand to closely grip the brush and thus better control its smoothing action. In addition, the shorter handle style is usually a wider or thicker handle, which fits a man's larger hand more comfortably and provides a stronger brush that won't break easily.

The Fuller Brush Company's laboratory manager, Bill Dayton, suggests another theory that explains why men's hairbrushes have gotten shorter and shorter over the centuries (many older men's brushes were indistinguishable from women's): "Men's brushes were designed to conserve space in military duffel bags and dop kits."

Submitted by Anne Taylor Spence of Washington, D.C.

Why Do Some Ladybugs Have Spots and Others Have None?

Ladybugs, beloved by children and politically correct animal lovers, have a great public relations person. In reality, they are just a type of beetle. There are somewhere between three thousand and four thousand species of ladybugs in the world.

Their colorings and markings vary so much that most entomologists have concluded that nothing in particular separates a ladybug with spots from a spotless individual. You can't tell the sex or gender of a ladybug from its markings. One individual ladybug from a given species might have spots; others of the same species, from the same region, might not.

The variations of coloring are almost endless. Ladybugs' bodies might be red with black markings, or orange with blue markings. Some have only two spots, while the Thirteen Spotted Lady Beetle, appropriately enough, sports thirteen. Why has nature provided them with such a seemingly random succession of markings?

Most entomologists believe that spots are there for defensive purposes. Robin Roche, keeper and entomologist at the Insect Zoo in San Francisco, told *Imponderables* that in nature, red and black (two of the most common colors for ladybugs' spots) are warning colorations. Other creatures, especially birds and rodents, learn that animals with certain colors sting or taste less than delectable. Even animals that don't emit toxins might mimic the appearance of spotted animals that do. Many ladybugs actually do make a lurking predator's life most unpleasant: Some spray blood, while others spray a poisonous fluid.

Although no one can be sure, most entomologists lend credence to the theory that regional spotting variations occur because the ladybugs' spots simulate the appearance of a more venomous animal. But others aren't so sure and wonder why, if the markings are so important to their defense, other individual ladybugs in the same region aren't born with a similar defense mechanism. Lynn S. Kimsey, director of the Bohart Museum of Entomology at the University of California, Davis, suggests that some variations in spotting might be due to temperature differences, "or it may be a genetic component of a population, much like the coloration of domestic dogs (e.g., dalmatians versus Labrador retrievers).

Submitted by Angel Vecchio of Fresno, California. Thanks also to Ashley Watts of Caledonia, Ontario.

Why Does a Loud Bang or Opening and Closing the Oven Door Sometimes Make Soufflés and Cakes Fall in the Oven?

Tom Lehmann, bakery assistance director at the American Institute of Baking, told *Imponderables* that while a cake is being baked, the batter rises to a point slightly higher than its fully baked height. The baking powder in the batter produces gas that causes the leavening effect. "At a time when the batter is at its maximum height, but has not 'set' due to starch gelatinization and protein coagulation, the batter is very unstable." The cake is at its most fragile and delicate because, according to bakery consultant Dr. Simon S. Jackel, "the air cells holding the entrapped gases are very thin and weak."

Not all cakes will crash if confronted with a loud noise. But most will fall during this vulnerable time during the cooking process, and soufflés are always in danger. Joe Andrews, publicity coordinator for Pillsbury Brands, explains:

> The basic structure of a soufflé is developed by egg proteins, which are whipped into a foam and then set by baking. When whipping of the

> egg whites occurs, large pockets of air are trapped by the albumen, and in the process, this protein is partially denatured. The denaturation (or setting) continues (along with the expansion of the air bubbles) when the proteins are heated in the oven. If the oven is opened while this expansion is taking place, the air pressure change and temperature change can cause the whole structure to collapse.

The most common bang, of course, is the opening and closing of the oven door. Anyone near a loudspeaker at a rock concert knows that sound vibrations can be felt; a soufflé or cake can be pummeled by a nearby noise. Although cakes are usually hardier than soufflés, Andrews indicates the same problems that afflict soufflés also make cakes fall,

> especially if the primary source of leavening for the cake is beaten egg whites (e.g., angel food or chiffon cakes). Layer cakes contain more flour and the structure is formed as much by starch gelatinization as egg denaturation, so they would not be as susceptible to falling when the door is opened—unless the door is opened too early in the baking process (during the first twenty minutes), before the cake structure has set.

Only when the internal temperature of the cake reaches a range of 160 to 180 degrees Fahrenheit is the cake out of the woods, because, as Jackel puts it, "the liquid batter is now converted to a solid cake structure."

Submitted by Sherry Grenier of Amos, Quebec.

The famous "Angel Food Cake" portal at Gateaux Cathedral.

Why Do Angel Food Cakes Have to Be Turned Upside-Down While Cooling?

As we just learned, angel food cakes are structurally delicate when baking, but once they've achieved a solid state, why in the heck do we have to turn them on their heads? We headed back to our trusty experts for their counsel. Tom Lehmann responded:

> Angel food cakes are really nothing more than an expanded egg white foam with sugar added for sweetness, and flour added to stabilize the foam and prevent it from collapsing during baking and cooling. Due to their inherent weakness, angel food cakes would collapse during baking and cooling if it weren't for two things. First, an angel food cake pan is never greased. This allows the batter to grip the pan sides for added support. The cakes are then stuck tightly enough to the pan after baking to allow them to be inverted without the cake falling out of the pan. By inverting the pan, we prevent the cake from further settling during cooling and obtain a light, tender finished cake.

Dr. Jackel notes that inverting the cake is absolutely essential for achieving an evenness of consistency:

> Although the top of the angel food cake has lost moisture in the oven during baking, and formed a dry skin, the bottom of the cake has retained some of the moisture and is slightly soft and sticky, because the bottom of the pan is not designed to release moisture as the cake bakes. When the cake is cooled, it is turned over so that the sticky, moist bottom of the cake has a chance to lose the extra moisture to the atmosphere and form a skin. The top has already formed the skin in the oven and therefore is already dry and firm.

Submitted by Gregg Hoover of Pueblo, Colorado.

Why Have Auto Manufacturers Moved the Brights/ Dimmer Switch from the Floorboard to the Stalk of the Steering Column?

We have fond memories of cross-country trips in which we were so bored during barren stretches that we would amuse ourselves by clicking the dimmer control on the floorboard, even though our lights weren't on. This may not compete with square dancing or coin collecting as a pastime, but it was some solace as we fantasized about the next odometer check or Stuckey's we might encounter.

Alas, our old diversion has now faded into nostalgia. In the 1970s, Detroit followed the lead of European and Japanese automakers and mounted brightness controls on a stalk of the steering column. At first we wondered whether this change was mandated by regulation, but we quickly learned there was no such requirement. An expert at the National Highway Traffic Safety Administration who prefers to remain anonymous informed us that the Department of Transportation only cares that there be a control to turn the high beams on and off, and that a (blue) light alerts the driver that the brights are on. (Red lights are reserved for warning indicators, such as overheating, oil shortages, etc.)

From the government's standpoint, the location of the control is not a safety issue, so the dimmer switch could be mounted on the ceiling and require a head butt to engage. So why did the automakers bother changing? We received five different explanations:

1. The move allowed auto manufacturers to put all the electrical features in the steering column instead of isolating one electrical element far away on the floor. This is why light, windshield wiper, and cruise controls have joined the horn and directional signals on the steering columns of most cars. Furthermore, as pointed out by Vann Wilber, director of safety and international technical affairs for the American Automobile Manufacturers Association, the floors of cars tend to get wet in the winter, and the water can seep into the electrical system, a potential safety hazard.

2. Consumers seem to prefer it. Wilber told *Imponderables* that the Big Three American automakers conducted human factors research indicating that drivers can identify hand-operated controls more quickly and easily than floor-mounted counterparts. Obviously, if the controls are adjacent to the steering wheel, the driver's hands are close to the beam control. Now that automobiles are often laden with as many gewgaws as jet instrument panels, the visibility of controls has become increasingly important. According to a member of the Society of Automotive Engineers' Lighting Committee, well-labeled stalk controls forestall drivers from looking around the dashboard, feeling around with their feet, or even worse, looking on the ground for the right pedal to depress, when they should be looking at the road.

Notwithstanding this reasoning, we must argue that markings on the stalk are of little use if the interior of the car is dark. Although one eventually becomes accustomed to the location of controls on one's own cars, it is disconcerting to rent an automobile and find oneself turning on the windshield wipers when one meant to cut off the high beams. Locating the old controls at worst required a one-time foot grope—knowing that there were no other controls on the floor made us less queasy about searching for it.

3. For people driving with standard transmissions, who must constantly use the left foot for the clutch, floor-mounted controls were often a nuisance and potentially even a safety hazard.

4. Mounting the brightness control on the steering stalk has enabled manufacturers to allow drivers to put on the high beams even when their lights

were not previously on. This feature makes it possible for drivers to alert the car in front of them to move over so that they can pass.

5. Because of the increase in international travel and alliances between American and foreign automakers, it makes sense to standardize as many features of automobiles as possible, particularly safety features.

We can't stop progress, we guess. But we're not happy about this particular change. If you are depressed about your car's barren floorboards, you may perk up a little when you find out that the issue of dimmer controls has inspired a joke among folks in the auto industry:

> General Motors is circulating a new service bulletin regarding cars with high beams on the stalk. G.M. is going back to the floor-mounted switch because too many_____(fill in favorite oppressed group) were getting their feet tangled in the steering wheel when they tried to turn on their brights.

And then again, maybe you won't perk up after hearing the joke.

Submitted by David Letterman, somewhere in Connecticut.

Why Is an Ineligible College Athlete Called a "Redshirt"? And Why Do Colleges Redshirt Players?

We were surprised at how difficult it was to obtain hard information about the history of redshirting. But every football source we contacted told us to contact Pat Harmon, legendary Cincinnati sportswriter and currently historian of the College Football of Fame. Harmon was kind enough to write us about the origins of this colorful term:

> At the University of Alabama many years ago, the coaching staff had recruited a large number of new students who were football players. Some of them were mature enough to work in the regular format—four years of college, four years of football.

> But if the coaches had an overabundance of player-candidates at one position—say tackle or end—they would decide to hold some of the newcomers back a year.
>
> These students would go to class for five years. *They would practice football for five years but play only four.*
>
> For that first year, when they practiced every day but were not used in games, they needed an identification so the coaches could spot them quickly. They were given red shirts [to separate them from the varsity playing squad].
>
> The practice of developing five-year players spread to other schools, and so did the use of red shirts. Thus a player who was held out for a year was called a redshirt.

The redshirted player lives in a twilight zone best described by writer Douglas Looney in a 1982 *Sports Illustrated* article:

> The redshirt gets to practice like the other players, gets chewed out like the other players, goes to sleep in meetings like the other players, and takes his lumps like the other players, except he doesn't get to play in games. Which is to say, he gets everything football has to offer but the fun.

College football researcher Ray Schmidt told *Imponderables* that in practice, many coaches and other players actually *do* take it a little easier on redshirted freshmen. After all, why should coaches "waste" their time trying to perfect a play with athletes who will never implement it?

The National Collegiate Athletic Association (NCAA) has had a long love-hate relationship with redshirting. The NCAA first legally adopted the practice in 1961 on behalf of a DePaul basketball player who did not play his freshman year. Because of technical regulations then in place, the player was free to play during the college season but ineligible during postseason competition.

Unfortunately, the redshirt rule was abused. Although technically legal, murmurs of discontent among coaches was heard when the head football coach of the University of Washington, Ray James, who had a particularly talented group of upperclassmen, redshirted twenty-one of his twenty-three freshmen in 1978. As Bob Carroll of the Pro Football Researchers Association told *Imponderables,* eventually "coaches started stashing players away simply to preserve their eligibility."

Of course, if the coaches' redshirting strategy works, the academic sophomore/football freshman starts off his actual intercollegiate play bigger, faster, and smarter than he would if he played right away. But redshirting can backfire. If he does not impress the coaching staff, he risks losing a scholarship for the next four years; if successful the freshman increases his potential marketability in the pros and could dominate his nonredshirted college competitors.

Submitted by Dr. John Nushy of Torrance, California.

If You Dig a Hole and Try to Plug the Hole with the Very Dirt You've Removed, Why Do You Never Have Enough Dirt to Refill the Hole?

After speaking to several agronomists, we can say one thing with certainty: Don't use the word "dirt" casually among soil experts. As Dr. Lee P. Grant of the University of Maryland's Agricultural Engineering Department remonstrated us, dirt is what one gets on one's clothes or sweeps off the floor. Francis D. Hole, professor emeritus of soil science and geography at the University of Wisconsin-Madison, was a little less gentle:

> What would you do if you were some fine, life-giving soil who is twenty thousand years the senior of the digger, and you were operated on by this fugitive human being with a blunt surgical instrument (but without a soil surgeon's license), and if you were addressed as so much "dirt," to boot? I am suggesting that a self-respecting soil would flee the spot and not be all there for you to manipulate back into the hole.

So there's the answer: The soil is offended by you calling it dirt, Loren, and has flown the scene of your crime against it.

We promised Grant and Hole we would treat soil with all the respect it was due, and temporarily suppress the use of the "d" word, if they would answer our question. They provided several explanations for why you might run out of soil when refilling a hole:

1. *Not saving all the soil.* Dr. Hole reported one instance, where in their excitement about their work, a team of soil scientists forgot to lay down the traditional canvas to collect the collected soil: "We had lost a lot of the soil in the forest floor, among dead branches and leaves."

2. *You changed the soil structure when you dug up the dirt.* Grant explains:

> Soil is composed of organic and inorganic material as well as air spaces and microorganisms. Soil has a structure which includes, among other things, pores (or air spaces) through which water and plant roots pass. Within the soil are worm, mole, and other tunnels and/or air spaces. All of this structure is destroyed during the digging process.

Hole confirms that stomping on the hole you are refilling can also compact the soil, removing pores and openings, resulting in plugging the hole too tight:

> It sounds like a case of poor surgery to me. You treated the patient (the soil) badly by pounding the wound that you made in the first place.

3. *Soil often dries during the digging/handling/moving.* Grant reports that the water in soil sometimes causes the soil to take up more space than it does when dry.

Both of our experts stressed that the scenario outlined by our correspondent is not always true. Sometimes, you may have *leftover* soil after refilling, as Dr. Hole explains:

> It is risky to say that "you never have enough soil to refill." Because sometimes you have too much soil. If you saved all your diggings on a canvas and put it all back, there could be so much soil that it would mound up, looking like a brown morning coffee cake where the hole had been.
>
> . . . you loosened the soil a lot when you dug it out. When you put the soil back, there were lots of gaps and pore spaces that weren't there before. It might take a year for the soil to settle back into its former state of togetherness. A steady, light rain might speed the process a little bit.

Submitted by Loren A. Larson of Orlando, Florida.

Why Was Twenty-one Chosen as the Age of Majority?

Has there ever existed a teenager who has not wailed, loudly and frequently, "Why do I have to wait until I'm twenty-one until I can (fill in the blank)?" To a kid with raging hormones, the number seems totally arbitrary.

And of course, the age *is* arbitrary. Now that some states have lowered the drinking age to eighteen ("If we are old enough to fight in Vietnam, we're old enough to vote and drink ourselves silly," went the argument), the number twenty-one seems downright capricious. How did this tradition begin?

Michael de L. Landon, professor of history at the University of Mississippi, provided us with the proper ammunition to blame the appropriate party: the British.

> Of course, twenty-one is approximately the age when both young men and women complete their full physical growth. More specifically, in medieval times in western Europe, young men of noble and knightly families normally left their homes to enter into service in the household of

someone of equal or higher rank (as compared to their parents) around the age of nine to eleven. Until fourteen, they served as pages, mostly under the supervision of the ladies of the household. From fourteen to twenty-one, they served as (e)squires attached to adult knights who, in return for having their horses attended to, their armor polished, etc., were supposed to train them in the knightly arts.

By the thirteenth century, twenty-one was customary age for a young man to be knighted. Likewise, among middle-class families in the towns and cities, a boy would normally be apprenticed at adolescence (i.e., around fourteen) to a "master" to learn a trade or craft. The customary apprenticeship period was seven years, until the age of twenty-one.

We wonder whether today's teenagers would exchange the right to drink at an earlier age for the right to leave home and work for up to twelve years before the age of twenty-one.

The English age of majority was by no means universal, even in Europe. Professor de L. Landon points out that in Roman law, children were "infants" until the age of seven; "pupils" until the age of puberty (girls, twelve; boys, fourteen); and minors until marriage for girls or age twenty-five for boys.

Submitted by Scott Wallace of Marion, Iowa. Thanks also to John Anthony Anella of South Bend, Indiana; and Joey Maraia of Nacogdoches, Texas.

Why Don't Disc Jockeys Identify the Titles and Artists of the Songs They Play?

We have a nifty secret for curing the morning blahs—sleep through them. Yes, we admit it: We're night people. We sleep until noon, run the shower, and flip the radio on to WHTZ, better known as Z100 in the New York City metropolitan area, and listen to the midday jock, Human Numan. Z100 is what the radio trade calls a CHR (Contemporary Hits Radio) station, a modern mutation of the old Top 40 format. Z100 has a small playlist of current songs.

Human's a terrific disc jockey. He's not full of himself. Doesn't reach for laughs. But we have one big complaint: He rarely, if ever, identifies songs. As we're writing this chapter, we've heard the new New Order single played at least twenty-five times on his show but have yet to hear the title identified.

Fate threw us into Human's lap one day, and we got to talk to him about this Imponderable. DJs have two options in identifying a song: introducing it before they play it, or "frontselling"; or playing the song and announcing the name of the recording artist and/or song afterward, or "backselling." The first thing that Human wanted to let *Imponderables* readers know is that the vast majority of DJs, especially in major urban markets, have little artistic control over what they play and what they say on the air. In a letter, Human discussed the pressures and constraints of a DJ in his kind of format and used a fifteen-minute segment of his show to demonstrate:

> Think of the DJs in the Top 100 markets as actors or football players. The coach designs the plays and the playwright gives the actor his lines: It's the same for the American DJ.
>
> The program director (PD) is the second most powerful person at a radio station, behind the general manager (GM). The PD hires the DJs and has the power to fire them, promote them, and has complete control over their shows.
>
> The PD creates a structure for the DJ's show called a *format clock*. This is a paper clock that has no hour hand because it is used every hour. On this clock, for example, it says where the one is: SEGUE, to proceed without pause (radio language for "shut up, just play the next song"). A DJ can never talk where the PD has indicated SEGUE on his routine clock.
>
> Then between the 1 and 2 on the clock, somewhere about seven minutes past the top of the hour, the PD might indicate LINER. This element means that the DJ has been given a 3" by 5" card with the "lines" he should ad lib or read verbatim, depending upon how strict the PD is. The LINER is a very important sell, one that the station must convey without any DJ clutter. The liner should not be diffused with additional information, such as a backsell of the previous record.
>
> The next element marked on the clock, at perhaps twelve minutes past, might say "BACKSELL/FRONTSELL NEW MUSIC—OPEN

SET." This element indicates that every hour at this point, the third or fourth record in the hour will be a brand new song that the PD wants to identify to the listener. Aha! The DJ may now ID the song. But notice he or she may ID only when indicated by the clock. Here my format clock also said "OPEN SET." This is the time when a DJ is free to express himself or herself (as long as the DJ remembered to sell the NEW MUSIC in this case!).

I'm just the tailback running up the left side, running a play the coach has called. I try to put my own spin on it, and dodge the tackles, but it is somebody on the sidelines calling the play.

PDs love to use a private Batphone setup in virtually every studio in every radio station. It's called the HOTLINE. If you don't follow the format, guess who's calling?

So if it's the PD calling the shots, why don't PDs instruct DJs to identify more songs? We talked to scores of disc jockeys and PDs and found absolutely no consensus about the wisdom of frequent song identification. Here are some of the most important reasons for lack of IDs, followed by the rebuttal case for more IDs.

1. *Research shows that listeners want more music and less talk.* Jay Gilbert, afternoon drive DJ on WEBN, Cincinnati, one of the first Album-Oriented Rock stations, told us that every research survey he has ever seen has indicated that most listeners want DJs to shut up and play more music. Originally, the relative lack of commercials and DJ chatter on FM helped the fledgling band win over AM listeners.

Sure, says Cleveland radio personality Danny Wright, who is generally against overdoing IDs, every poll he has seen in his twenty years in broadcasting indicates that listeners hate jocks who talk too much. But then who are the most popular people on the air? According to Wright, "the folks with the oral trots"—Rush Limbaugh, Howard Stern, Rick Dees, Scott Shannon, etc. Wright believes that if a jock has nothing to say, he is better off just playing music, but that audiences love patter if it is entertaining.

2. *IDs slow down the show.* In order to speed up the pace of the show and to provide the illusion of more music being played, stations will do everything from playing records at a higher than normal speed to instructing DJs to talk over the music. To many PDs, back announcing, in particular, is just dead air, particularly when the time could be devoted to more jingles promoting the call letters of the station.

Of course, the five or ten seconds devoted to identifying a song could be spent playing more music, but then perhaps a radio show should be more than a jukebox with commercials. Al Brock, a PD and on-air personality at WKLX, an oldies station in Rochester, New York, told *Imponderables* that identifying a song is a way of connecting the DJs with the music, showing listeners that the jocks are interested in and committed to the music. PDs who are for frequent IDs see them as part of the music programming, while anti-ID PDs see them as part of the talk. Brock feels strongly enough about the issue to try to frontsell or backsell every song on the station (which can't always be done, because of time constraints).

3. *Why tell audiences what they already know?* A classical music station usually IDs every selection it plays, because the audience might not be able to recognize a particular piece or the conductor and orchestra. But does a DJ really have to tell an audience "That was Whitney Houston and 'I Will Always Love You'?"

The answer of the pro-ID side is, "Yes, you do." Al Brock informed us that most people know some songs by titles and other by artists but that few can remember both. For example, after the Righteous Brothers' "Unchained Melody" was rereleased during the popular run of the movie *Ghost*, the song was not only played on oldies stations (it never stopped being played there) but promoted as if it were a new song on many CHR and AC (Adult Contemporary) stations. Yet listeners constantly called to ask the name of the song or the group who sang it. Another DJ told us that every time he plays Paul Stookey's "The Wedding Song" (the title is not part of the lyrics), even if he front- or backsells it, he gets calls asking, "What song was that?"

Obviously, the need for IDs depends upon the format of the station and the familiarity of a given song. Virtually every PD and DJ we spoke to identified a brand new song, one that the station has been playing for two to four weeks. (These songs are called "currents.") All agreed that the songs least needed to be ID'd are songs that are no longer current but are still popular and haven't left the playlist. These are known as "recurrents" and are usually played less than "currents" but more than oldies. Some PDs argue that oldies don't require IDs because they are so familiar, but even this strategy has pitfalls, for oldies stations are trying to attract younger listeners, including people who might not have been *alive* when a song was recorded.

4. *IDs create clutter.* An old broadcasting bromide is that each music set on a radio show should stress one thought. Considering that there are many elements in a radio show—music, talk, promos, ads, weather, contests, jingles—IDs can cause more confusion than enlightenment. Steve Warren, a veteran New York radio personality now heading his own programming consulting company, MOR Media, reminded us that for most of the audience, radio is a secondary medium. Most listeners are doing other things, such as driving cars, sewing, or taking a shower, while listening to the radio. Overloading any format with too much information can backfire.

Even pro-ID programmers realize that, for example, during morning drive shows, when information about weather and traffic may be paramount and commercials are most frequent, backselling may not be prudent. They often fine-tune their volume of ID's by daypart.

5. *IDs slow the momentum of the show.* One of the tenets of CHR radio is "always move forward." The name of the ratings game in radio is to keep listeners as long as possible. Unlike television, where viewers generally have some loyalty to particular shows and are likely to stick with them for the half-hour or hour, PDs are acutely aware that listeners in automobiles have push-buttons that can "eject" their station the moment they hear an unwanted song or one too many commercials. This is one reason why many stations start a new song before the DJ talks over it—subliminally, this tells the listener, "Don't worry, there is no advertisement coming up."

One of the main strategies for keeping us tuned in longer is to promote what is coming up next. As Danny Wright puts it,

> Never talk about last night or a movie you saw last week or what you just played. Billboard the next few tunes and events to keep listeners sticking around.

PDs employing this strategy often frontsell. Before a commercial break, a jock might say, "Coming up, the new Eric Clapton, Whitney Houston, and an oldie by the Beatles." The hope is that the listener will stay glued to the station if she likes one or more of the songs.

Of course this strategy can backfire too. If a listener would rather hear fingernails on a blackboard than Whitney Houston, he may desert the station, even if he was mildly curious about the identity of the Beatles oldie.

Many of the "more music, less talk" stations feature "music sweeps," in which five or more songs are played in a row without commercial interruption. Frontselling eight songs at a time is tedious, and backselling is deadly. Some stations solve the problem by frontselling only one or two songs and doing the same on the back end. Some feature what Al Brock calls "segue assists," in which the jock IDs the song before or after every record.

> 6. *Selling records isn't a radio station's job.* We spoke to several radio programmers who echoed this sentiment. The trade association of the recording industry, the Recording Industry Association of America (RIAA), launched a campaign to promote IDs, plastering stickers on DJ record copies saying, "When You Play It, Say It," the "It" meaning title and artist. In 1988, the RIAA released a study of over one thousand radio listeners, between the ages of twelve and forty-nine, indicating that about two thirds of the respondents would like more information about the records they heard on radio. Listeners between twenty-five and forty-nine years old were particularly vehement, and several programmers we spoke to revealed that the lack of IDs has surpassed "too much talk, not enough music" as the number one complaint of listeners.

Increasingly, radio stations are conducting "outcall" research, telephoning listeners and asking them about their musical preferences. This type of research is of little value if respondents don't know the titles and artists of the songs played on the stations. One PD we consulted, who wished to remain anonymous, indicated that his policy of heavy backselling had nothing to do with helping record companies:

> We try to backsell as much as possible for two reasons. First, it answers the listeners' primary question: What was that we just heard? Second, it helps us with our research. How are we supposed to ask listeners to call in our request line if they don't know what they've heard on our station?

Consultant Steve Warren suggests that there *are* alternate ways of supplying listeners with information about titles and artists, including manned request lines and listener hotlines (in which an employee answers questions about the music, the station, contests, etc.). Warren indicated that at times it doesn't hurt to have calls come in directly to the DJ—it's a good way for jocks to stay in touch with their fans.

Disc jockeys have so many chores to perform besides listening to music that many are understandably not excited about IDs; after all, their time on the air is extremely limited. So, we guess we can't be too hard on our very Human Numan for not frontselling or backselling every song. After all, he estimates that on his average three-hour shift, he speaks on-air for a grand total of *seven minutes.*

Submitted by the guy in the shower, New York, New York.

Why Do Pigeons Make a Whistling Sound When They Take Off in Flight?

Those aren't pigeons' voices but rather their wings you are hearing. Bob Phillips, of the American Racing Pigeon Union, told *Imponderables* that we are hearing the sound of air passing through feathers that are spread wide for acceleration, beating faster for lift, and spread wide for takeoff. Although we tend to associate this kind of high-frequency noise with hummingbirds, many birds produce similar tones, not unlike the sound of the wind whistling through the branches of trees.

Submitted by Martin C. Farfsing of Redwood City, California.

To prove the wholesome, family orientation of our readership, we can point to a surprising cluster of Imponderable obsessions about the subject of milk. Perhaps not the sexiest topic, but certainly among the most nutritious.

What's the Difference Between Skim and Nonfat Milk? And How Do They Skim the Fat from Whole Milk?

Don't believe it if anyone tells you there is any difference whatsoever. By law, skim milk and nonfat are the same: containing less than 0.5 milkfat content. (The milk solids that are *not* fat must equal or exceed 8.25 percent.) In practice, all the fat possible is eliminated from the product.

Any nonfat (or lowfat) milk that is shipped interstate must contain added vitamin A. Most of the vitamin A content in milk is contained in the milkfat. Most manufacturers add enough to equal the amount of vitamin found naturally in whole milk.

How do they separate the fat from whole milk? Our favorite dairy consultant Bruce Snow, explains:

> When milk comes from old Bossy, it contains somewhere between 3 to 4 percent butterfat content (sometimes a percentage point more from cows like Guernseys and Jerseys). To obtain true skim milk, a machine called a "separator" is used. It whirls the milk around, and because the fluid and the butterfat content have different weights, centrifugal force separates the two ingredients into skim milk and cream. The cream is used to make butter, ice cream, whipped cream, etc.

Submitted by Herbert Kraut of Forest Hills, New York.

Why Does Some Lowfat Milk Contain One Percent Fat and Other Lowfat Milk Contain Two Percent Fat?

As we just learned, most cows naturally produce milk containing from 3 to 5 percent butterfat. In most states, "whole" milk is defined as milk with at least 3.25 percent butterfat. Lowfat milk, then, is any milk that falls between .5 percent (skim or nonfat) and 3.5 percent (whole).

In practice, 1 and 2 percent milks are the most popular types of lowfat milk. In fact, we've never seen 3 percent milk, probably because that one half a percent would not reduce the calorie count enough to appeal to dieters.

Many consumers were sick of looking at what looked like water residue on the bottom of their cereal bowls; lowfat has been steadily gained market share for the last thirty years, stealing customers from both former skim milk and whole milk drinkers. In fact, lowfat milks outsell "whole" milk in most parts of the country. Two percent seems to be winning the cash register battle against 1 percent, but not without a cost to the waistline; that extra percent of fat adds about thirty calories to each cup of 2 percent lowfat milk.

Submitted by Herbert Kraut of Forest Hills, New York.

Why Do Plastic Gallon Milk Containers Have the Counter-Sunk Dips on Their Sides?

According to Michelle Mueckenhoff, technical services manager of the Dairy Council of Wisconsin, and every other dairy expert we bored with this question, those dips are there to provide structural support and strength to the container. And nothing more.

Submitted by Daniell Bull of Alexandria, Virginia.

Why Aren't "Green Cards" Green?

We've never been ones to make cheap, easy jokes about our federal government. Sure, there is excess and incompetence in any large conglomeration of workers. We were confident that there was a perfectly brilliant strategy behind naming what are most often blue cards "green cards."

So we contacted the Immigration and Naturalization Service and were lucky enough to come in contact with Elizabeth A. Berrio, chief of the INS Historical Reference Library, who specially prepared a document to share with *Imponderables* readers. And we're happy to conclude that there is a totally logical reason why green cards aren't green. Well, would you believe semilogical?

> What we know as a "green card" came in a variety of different colors at different times in its history. We still refer to them as "green cards"

for the same reason dismissal notices are called "pink slips," sensationalized news is called "yellow journalism," and intended distractions are called "red herrings." In each case, an idea was originally associated with an actual item of the respective color. A Lawful Permanent Resident (LPR) alien living in the United States may carry a card that is not green, but refers to it as a "green card." The alien does so because the card bestows benefits, and those benefits came into being *at a time when the card was actually green.*

The green card is formally known as the Alien Registration Receipt Card, form I-151 or I-551. The first receipt cards were form AR-3 and were printed on white paper.

This receipt proved that a noncitizen of the United States did register, but it didn't indicate whether the alien was legal or illegal. After World War II, when a new wave of immigration began, the INS started issuing different documents to indicate whether an alien was a visitor, a temporary resident, or a permanent resident.

> This method . . . helped to identify the immigration status of each alien. Thus, the small, green I-151 had immediate value in identifying its holder as a LPR, entitled to live and work indefinitely in the United States. As early as 1947, LPRs protested delays in processing their I-151s, complaining that employers would not hire them until they could prove their permanent resident status.

In 1951, the green cards became even more valuable:

> regulations allowed those holding AR-3 cards to have them replaced with a new form I-151 (the green card) . . . only aliens with legal status could have their AR-3 replaced with an I-151. Aliens who applied for replacement cards but could not prove their legal admission into the United States, and for whom the INS had no record of legal admission, did not qualify for LPR status and might even be subject to prosecution for violation of U.S. immigration laws.
>
> By 1951, then, the green Alien Registration Receipt Card Form I-151 represented security to its holder. It indicated the right to permanently live and work in the United States and instantly communicated that right to law enforcement officials. As a result of the card's cumbersome official title, aliens, immigration attorneys, and enforcement officers came to refer to it by its color. The term "green card" designated not only

the document itself, but *also the official status* desired by so many legal nonimmigrants (students, tourists, temporary workers) and undocumented (illegal) aliens. The status became so desirable that counterfeit form I-151s became a serious problem.

To combat document fraud, the INS issued nineteen different designs of the I-151 between its introduction in the 1940s and its complete revision in 1977. One alteration to the design in 1964 was to change the color of the card to blue. The 1964 edition was a pale blue. After 1965, it was a dark blue. Regardless of color, the I-151 still carried with it the benefits indicated by the term "green card," and those who wanted, obtained, issued, or inspected I-151s continued to refer to it by that name.

The INS has not given up on foiling counterfeiters. In 1977, it issued a machine-readable receipt card and keeps experimenting with new colors, including such unlikely choices as pink and pink-and-blue. But Berrio is resigned: "Despite these changes in form number, design, and color, the INS document which represents an alien's right to live and work in the United States will probably always be known as a "green card."

Submitted by Eileen Joyce of Texarkana, Texas.

When Do Kids Turn into Goats?

When they have their bar mitzvah?

But seriously, folks, though it may have been submitted by a "kid" in Mary Helen Freeman's Aiken, South Carolina, Millbrook Elementary School class, this Imponderable was sufficient to stump most of the goat authorities we contacted. For example, John Howland, secretary-treasurer of the American Goat Society, was modest enough to admit he wasn't sure of the answer and was kind enough to write to several other experts. They couldn't reach a definitive consensus: Some argued for four months; others for six months; and some maintained that kids didn't turn into goats until they were old and large enough to breed.

Rowland then consulted the text *Raising Milk Goats the Modern Way,* by Jerry Belanger:

> Mr. Belanger said that kids are ready to breed when they are about seven months old and weigh about seventy-five to eighty pounds. In his glossary, he says: "Kid: a goat under one year of age."

Bonnie Kempe, secretary of Alpines International, concurred with Rowland that definitions of "kid" vary from expert to expert, but she did offer what she thought was the most popular definition:

> Baby goats are called "kids" the first year of their life. The second year they are called "yearlings," and after age two either "does" or "bucks."

Still, some veterinarians we spoke to felt that once a goat can breed and bear offspring, it is inappropriate to call mama or papa a kid.

Submitted by Ivy Moore of Aiken, South Carolina.

Why Do Baked Hams Usually Have a Checkerboard Pattern Along the Top?

Most of the time, the checkerboard pattern is created by the cook scoring the top of the ham for purely decorative reasons. Unlike the brittle skin of a turkey, it is quite easy to cut superficial slices in ham, and many cooks find the pattern visually pleasing.

Chances are, though, that scoring of ham was originally initiated for more practical reasons. According to pork experts at the National Live Stock & Meat Board,

> . . . this process probably began with the old-style hams that had more fat. The scoring, or slicing, of the surface may have been done as a way to allow the fat to drain during cooking. Hams today are much leaner, so the scoring may be done simply for visual reasons.
>
> Since the hams are so lean, it is important not to score too deep. This will cause the natural juices to run out and make the ham very dry.

Robin Kline, director of consumer affairs at the Pork Information Bureau of the National Pork Producers Council, concurs with the "decorative" theory and adds that many other decorations are used routinely to embellish the cooked ham:

> One might also stud the top of the ham with cloves. You've probably seen pineapple rings, maraschino cherries, and pecan halves. Different strokes . . .

Occasionally you will buy a ham with a checkerboard pattern already emblazoned in the animal. These are the imprints of the netting used to hold and shape the meat during curing. The nets may be made of rubber-elastic, plastics, or natural fibers. According to Anne Tantum, of the American Association of Meat Processors, nettings (also known as "stockings") are used particularly often in the curing of boneless hams, which tend to bulge if left to cure without "undergarments."

Ham stockings come in many configurations, but most often the resultant patterns are square, rectangular, or diamond-shaped. And although these designs were probably the last thing on the minds of the ham processors, net-created patterns save the cook the not particularly time-consuming task of scoring the ham to create a little ocular razzle-dazzle.

Submitted by Wayne Rhodes of Deerfield, Illinois.

What Is the Emblem on the Pittsburgh Steelers' Helmets? And Is There Any Particular Reason Why the Pittsburgh Steelers Are the Only NFL Team to Have Their Logo on Only One Side of Their Helmets?

We thought this Imponderable might be a little obscure to include here, but when we found out that the Pittsburgh Steelers public relations department developed a form letter expressly to answer it, we realized that football fans must be burning to know all about the Steelers' helmet emblem. So here's the form letter:

> The emblem, called a steelmark, was adopted in 1963 and is the symbol of the Iron and Steel Institute. There is not a special reason as to why the emblem is only on the right side. That is the way the logo was originally applied to the helmet, and it has never been changed.

So many NFL teams redo their helmet design at the drop of a hat, so to speak, that our guess is that in 1963, the Steelers were not alone in their single-sided emblem configuration.

Submitted by Sue Makowski of Depew, New York. Thanks also to Thomas Ciampaglia of Lyndhurst, New York.

Why Did the Rabbit Die When a Pregnant Woman Took the "Rabbit Test"?

Ever since we were babes (as in "babes in the woods," not as in "hot babes," of course), popular culture, especially bad jokes, has informed us that "the rabbit died" meant "pregnant." But we always wondered why a rabbit had to die in order to diagnose a pregnancy. So we were gratified when this Imponderable was sent in by a reader who happens to be a physician. If he didn't know, maybe we weren't so dim-witted for not knowing ourselves.

At its height of popularity, the rabbit test would be administered to women after they missed two consecutive menstrual periods. A small sample of urine was injected into a female rabbit. But why urine? Why a rabbit?

Urine has been used to diagnose pregnancies as far back as the fourteenth century B.C. by the Egyptians. They poured urine on separate bags of barley and wheat. If either grain germinated, the woman was pregnant. They believed that if the wheat germinated, it would be a boy; the barley, a girl. There were probably a lot of unused cribs and miscolored baby clothing in ancient Egypt.

The early Greek physicians also dabbled in urine analysis for the detection of pregnancy. In his book *Obstetric and Gynecologic Milestones*, Dr. Harold Speert notes that urine analysis was a particular favorite of medieval English quacks, often called "piss-prophets," who claimed to diagnose just about any malady from indigestion to heartache. Reaction against these charlatans was so strong that urinary diagnosis was rejected during most of the eighteenth and nineteenth centuries by reputable physicians.

But in 1928, two German gynecologic endocrinologists, Selmar Ascheim and Bernhard Zondek, announced a urinary test that could be replicated easily throughout the world. They injected urine into five infant mice. Ascheim and Zondek explained why they needed five mice:

> Five infantile mice are used for each urine examination. The urine must be tested on several mice because an animal may die from the injection, but more important because not all animals react alike. . . . *The pregnancy reaction is positive if it is positive in only one animal and negative in the others.*

The A–Z test, as it has become known, is still the basis for all urine-based pregnancy exams, including the rabbit test.

So why the switch from mice to rabbits? Dr. T.E. Reed, of the American Rabbit Breeder's Association, explained the advantages of rabbits, and we promise to get through this discussion with no cheap "breeding like rabbits" jokes.

Most mammals have "heat" cycles, when females are receptive to the male. These cycles are physiologically based and are accompanied by changes in hormonal levels. The ovary is affected by the estrogenic hormone, the animal ovulates, and then is receptive to the male for breeding.

But the domestic rabbit is different, as Reed explains:

> The rabbit does not ovulate until it has been mated with the buck. The rabbit then ovulates ten hours later and the sperm that was deposited during the copulation process will fertilize the ovum.

> The uniqueness of domestic rabbits' physiology of reproduction is what allowed the pregnancy tests for humans to be utilized. Virgin does were used in the "rabbit test." Because researchers used does that had not been mated, the ovaries of the animal had never produced follicles from the ovaries.

Rabbit tests proved to be faster and more reliable than the original A–Z test.

But why did a pregnant woman's urine kill the rabbits? Ah, the nasty little secret: The test itself did not kill the rabbits, as Reed explains:

> The rabbit does not die of natural causes. The rabbit is euthanized after a specific amount of time [usually forty-eight hours after the first injection] has passed after being inoculated and the ovaries observed by the diagnostician. When the woman is pregnant, the follicles, which look like blisters on the ovary, would be present. If the woman was not pregnant, the ovary would be smooth as in virgin does.

The inventor of the rabbit test, Maurice Harold Friedman, injected the rabbits three times a day for two days, but later practitioners simplified the procedure to one injection and a twenty-four-hour waiting period. Through trial and error, researchers later found that it was not necessary to kill the rabbit at all, and one rabbit was used for several tests, after allowing the ovaries to regress after a positive result.

Although the theory behind the rabbit test was perfectly sound, one problem in reliability persisted: The rabbits chosen weren't always virgins, resulting in false positives. More sophisticated tests were developed without needing animals at all. But even modern laboratories, like the home pregnancy kits, measure the same hormone levels that Friedman, Ascheim and Zondek, and maybe even the piss-prophets and ancient Egyptians predicted pregnancy by.

Submitted by Dr. Ray Watson of Shively, Kentucky.

Why Is It That What Looks to Us Like a Half-Moon Is Called a Quarter-Moon by Astronomers?

An intriguing Imponderable, we thought, at least until Robert Burnham, editor of *Astronomy*, batted it away with the comment, "Aw, c'mon, you picked an *easy* one this time!"

Much to our surprise, when astronomers throw lunar fractions around, they are referring to the orbiting cycle of the Moon, not its appearance to us. *Sky & Telescope*'s associate editor, Alan M. MacRobert, explains:

> The Moon is *half* lit when it is a quarter of the way around its orbit. The count begins when the moon is in the vicinity of the sun (at "new Moon" phase). "First quarter" is when the Moon has traveled one-quarter of the way around the sky from there. The Moon is full when it is halfway around the sky, and at "third quarter" or "last quarter" when it's three-quarters of the way around its orbit.

Robert Burnham adds that "quarter-Moons" and "half-Moons" aren't the only commonly misnamed lunar apparitions. Laymen often call the crescent moon hanging low in the evening sky a "New Moon," but Burnham points out that at this point, the moon is far from new: "In fact, by then the crescent Moon is some three or four days past the actual moment of New Moon, which is the instant when the center of the Moon passes between the Earth and Sun."

Submitted by Susan Peters of Escondido, California. Thanks to Gil Gross, of New York, New York.

What's the Deal with the Grades of Architectural and Art Pencils? What Do "H," "HB," "F," "B," and "E" Stand For?

Here's the code: "H" stands for Hardness and "B" stands for "Black." With pencils, "hard" means a pencil that yields a lighter image. "Soft" pencils provide darker images. In this case, "black" means soft.

From hardest to softest, these are the grades: 9H, 8H, 7H, 6H, 5H, 4H, 3H, 2H, H, HB, F, B, 2B, 3B, 4B, 5B, 6B.

HB is the equivalent of the "regular" number 2 pencil. The grade "F" is a grade between "HB" (hard and black) and "B." Ellen Carson, of pencil manufacturer Empire Berol USA, told us that "F" was originally introduced for taking shorthand, because it was "hard enough to withstand extended use without resharpening and black enough so as to ensure that the shorthand was subsequently legible."

The "E" grades are used to designate the hardness of Filmograph leads. According to Carson, these leads are produced for plastic film and used in technical drawing. Filmograph leads contain no graphite and are based upon polymers and carbon black: "They are used in order to prevent the written or drawn line from being smudged when the drawing is being handled." E1 is the softest grade and E4 the hardest.

Submitted by Myron Kozman of Webster Groves, Missouri. Thanks also to Carol McDaniel of Castro Valley, California, and Roslyn J. Dy of Charleston, West Virginia.

How Do Hermit Crabs "Relieve Themselves" When in the Shell Without Getting Filthy? Or Do They Get Filthy?

The last few chapters have been altogether too pleasant. We don't want you to get complacent; it's time for a real gross-out now.

We often hear the bromide about the perfection of Mother Nature—that everything is part of her plan. But she had some strange plans in mind when she created the hermit crab. A "hermit crab" is not a particular species or family of crab; the term refers to various crabs that have soft abdomens and live in the empty shells of mollusks. The vision of a crab living in a vacated snail shell isn't too appetizing to begin with; but combine it with the crustacean's questionable bathroom habits, and we are stuck with one unpleasant visual image.

We asked one of our favorite crustacean experts, Dr. Darryl Felder, chairman of the biology department at the University of Louisiana, Lafayette, to explain our Imponderable. Even this expert couldn't remain totally clinical:

> Their urine is passed from pores at the base of the antennae, near the base of the eyes (as in all crabs, shrimp, and lobsters). The fecal material is passed from the posterior end deep within the shell and moved as a string-like fecal strand out the aperture of the shell. You too would be a "hermit" if you did such things in your home.

Not content to render an expert opinion on such a weighty subject alone, Dr. Felder was kind enough to pass along our query to Dr. Rafael Lemaitre, a research zoologist at the National Museum of Natural History at the Smithsonian Institution. Dr. Lemaitre concurred with all of Dr. Felder's conclusions but was kind enough to provide even more repellent details:

> . . . many worms, amphipods [crustaceans with one set of feet for jumping or walking and a separate set for swimming], and other animals are frequently found inside shells inhabited by hermits. It is quite possible that some of these "associates" help in the recycling of the hermit's wastes.

Lemaitre adds that the hermit crab can, by itself, create the currents necessary to flush the wastes out of its system. And if its plumbing fails, and it gets too grubby for the crustacean, it can always do what is most characteristic of a hermit crab: Ditch the shell it's living in and find more pristine accommodations elsewhere.

Submitted by Elaine Coyne of Brick, New Jersey.

Why Does Getting a Hair in Our Mouth Make Us Gag?

Our correspondent, Ilona Savastano, was passionate in her need for an answer to this burning Imponderable:

> How are we able to swallow just about any type of food, at times very large mouthfuls of it, with no problem, and yet we nearly gag to death if we get a tiny little hair in our mouth? I feel even one piece of fur in our mouth doesn't quite bring the same "yucky" reaction.

We flew your question by our dental experts, who are used to patients gagging (sometimes even *before* they receive their bill).

Their first reaction was to emphasize the sensitivity of the mouth. Dentist Ike House, of Haughton, Louisiana, amplifies:

> Our mouths are the most sensitive parts of our bodies, especially at birth and in childhood. Children use their mouths for food (nursing), comfort (thumb-sucking or pacifier), pleasure (witness the random exploration of children with their hands in their mouths), and exploration (children put a foreign object in their mouths to determine what it is). Because of our pattern of oral stimulation and exploration from an early age, the mouth is very sensitive.

Now that we have established how sensitive our mouths are, we might ask whether there is anything particularly nasty about the hair itself that might cause particular problems if found in the mouth. Yes, indeed, insist our experts. Dentist Barnet Orenstein, of the New York University David B. Kriser Dental Center, explains:

> Physically, a hair has two sharp ends capable of stimulating the very sensitive mucous membrane lining the oral cavity. Furthermore, the fine diameter and convoluted shape of a hair enable it to adhere to the mucosa. Dislodgement with the tip of the tongue is virtually impossible.

The gag reflex isn't necessarily more likely to occur when the mouth is full than when it contains one lonely hair. We can be tickled more easily by the light touch of a feather on the neck than by a hard rubbing of a bulkier object.

But the main culprit in hair-gagging is well above the throat. All of our dental experts think that the main cause of hair-gagging is psychological.

Dr. House reports that he has often been able to decrease or eliminate a patient's gag reflex simply by talking to him or her about the problem. And he opines that the nature of a foreign body determines our reaction to it:

> For example, spaghetti would not feel much different in our mouth than worms (assuming you had some GREAT sauce, and the worms were already dead and not wiggling around) but most people would choke at the thought. Hair is perceived as being dirty by most folks, witness the displeasure people have with finding a hair in their food. During lovemaking, however, touching your partner's hair with your mouth might be enjoyable.

So buck up, Ilona. Sure, you may not be able to control your gag on a single strand of hair, but blame the messenger, not the message. Gagging can be good for you, as Dr. Orenstein explains:

> The gag reflex is one of the many defense mechanisms nature has so miraculously endowed us with. Even infants will react violently to tickling of the soft palate. If it were not for this mechanism, many of us would expire by having the airway shut off by some foreign body lodged in our throat.

Submitted by Ilona M. Savastano of Cleveland, Ohio. Thanks also to Herbert Biern of Reston, Virginia.

Why Does Pasta Create Foam When Boiling?

Pasta is made from durum wheat, a particularly hard wheat. More precisely, pasta is created from durum wheat *semolina,* fine particles derived from the much coarser durum. The extraction of the semolina is largely responsible for the foaming of pasta when cooking, as Farook Taufiq, vice-president of quality assurance at Prince Company, explains:

> Durum wheat semolina consists of carbohydrates (starches) and protein. In the process of grinding wheat to extract semolina, some starch links are broken.
>
> When pasta is put in boiling water, these broken starch links swell up, taking in tiny air bubbles, along with water. These air bubbles come to the surface of the boiling water and appear as foam. So the foam is a combination of starch molecules, water, and air.

Submitted by Sam Rosenthall of Amherst, Massachusetts.

Why Do Many Elderly People, Especially Those Missing Teeth, Constantly Display a Chewing Motion?

Dr. John Rutkauskas, of the American Society for Geriatric Dentistry, consulted with two of his geriatric dentistry colleagues, Dr. Saul Kamen and Dr. Barry Ceridan, and told *Imponderables* that this chewing motion is found almost exclusively in people who have lost teeth. On rare occasions, certain tranquilizers or antidepressants (in the phenothiazine family) may cause a side effect called tardive dyskinesia, an inability to control what are ordinarily voluntary movements. These movements are as likely to involve the nose as the mouth or jaws, though.

In most cases, Rutkauskas believes that the chewing motion is a neuromuscular response to the lack of teeth: an attempt by the oral cavity to achieve some form of equilibrium. In particular, these sufferers can't position their upper and lower jaws properly. With a full set of ivories, the teeth act as a stop to keep the jaws in place.

Of course, most people who lose teeth attempt to remedy the problem by wearing dentures. And most people adapt well. But Ike House, a Louisiana dentist and *Imponderables* reader (we're sure he is prouder of the first qualification), told us that a significant number of elderly people have lost the ability to wear dentures at all because of an excessive loss of bone:

> They can close their mouth much fuller than they would with teeth present, resulting in the "nose touching skin" appearance of many elderly folks. Since the normal "rest position" of about 2–3 mm between the upper and lower natural or artificial teeth is not able to be referenced, they may be constantly searching for this position.

Many elderly people who wear dentures feel that the prostheses just don't feel normal. And restlessness leads to "chewing in the air," as House amplifies:

> If you had two objects in your hands, such as two pecans or two coins, you would probably manipulate them in some way. When not using

a pen or pencil, for example, but holding it passively, we usually move it in our hand. It may be that folks wearing dentures constantly manipulate them in some way just because objects being held but not used are often moved by unconscious habit.

I have a great-uncle who lets his upper teeth fall down between words and pushes them back up against his palate. This is a most disconcerting habit to his family! I know some elderly patients cannot tolerate dentures in their mouths unless they are eating because they can't leave them alone.

Barnet B. Orenstein, an associate clinical professor of dentistry at New York University's College of Dentistry, told *Imponderables* that the tongue is often the culprit in creating the chewing motion:

> Elderly people often display a constant chewing motion because, having lost their lower teeth, their tongue is no longer confined to the space within the dental arch. The tongue spreads out and actually increases in size. What appears to be a chewing motion is actually a subconscious effort to find a place for the tongue.

The last time we were at the *Imponderables* staff's official dentist, Phil Klein, of Brooklyn Heights, we asked him to wrestle with this mystery while he mauled our molars (and we pondered whether we could deduct the office visit from our income tax as a research as well as medical expense). Much to our relief, Dr. Klein concurred with the theories stated above but raised the possibility of a few others, including rare neurological conditions and grinding of teeth to the point where the lower and upper jaws can't mesh comfortably.

Klein also mentioned that problems with salivation, and particularly dryness, is a constant problem for numerous elderly people, and many with this problem move their mouth and jaw in response to this dryness.

And then he told us we had no cavities.

Submitted by Dennis Kingsley of Goodrich, Michigan.

Note to IRS: We deducted our trip to Dr. Klein as a medical expense.

Do Butterflies (and Other Insects) Sneeze or Cough? If So, Do They Do So Loud Enough for Humans to Hear?

All of the entomologists we contacted were sure about this Imponderable. Butterflies and other insects don't sneeze or cough. It's particularly difficult for them to sneeze, as they don't possess true noses.

So then how do insects breathe? Leslie Saul, Insect Zoo director at the San Francisco Zoo, explains:

> Butterflies and other insects breathe through holes in the sides of their bodies called spiracles. Spiracles are provided with valvelike devices that keep out dust and water. Some insects, such as some flies, june beetles, lubber grasshoppers, and notably the Madagascar hissing cockroaches, make sounds for communication purposes by forcing air out through the spiracles. Hence they hiss.

Karen Yoder, certification manager of the Entomological Society of America, concurs with Saul and adds that it isn't possible to hear insects breathing with the naked ear, either:

> In my days of insect appreciation, I have never heard the expiration of air from an insect. . . . Certainly, it could be possible to hear the transpiration in insects with the aid of an amplifier, but not with the naked ear.

But one needn't be wearing a stethoscope to hear the aforementioned hissing cockroach, better known to entomologists as *Gromphadorhina portentosa.* Our trusted informant, Randy Morgan, head keeper at Cincinnati's Insectarium, wrote an article for *Backyard Bugwatching,* a favorite magazine of our family's to pass around at barbecues, in which he chronicles the decibel-producing potential of these two- to three-inch cockroaches.

The Madagascar hissing cockroaches produce the hiss by contracting their abdomens and pushing the air out of constricted spiracles: The noise can be heard from several feet away. Whereas most cockroaches deter predators by running away, flying away, or producing unpleasant secretions, not so the *Gromphadorhina portentosa:* "Their secretive nature and ability to hiss seem to be their primary defense against enemies." Morgan cites an example of a lemur, eager to dine on our poor cockroach. But the hiss convinces the lemur it might have a rattlesnake or other dangerous critter on its hands: "In the leaves, a hissing cockroach continued feeding, unaware it had just narrowly escaped being eaten." Even if the cockroach wasn't aware of its near demise, the lemur's flight wasn't a coincidence. The hiss is a voluntary reflex, generally used only when a cockroach is in danger from predators or competing for mates.

Submitted by Marti Miller of Flagstaff, Arizona. Thanks also to Russell Shaw of Marietta, Georgia.

What Is the Liquid That Forms on Top of Yogurt? Is It Water or Does It Have Nutrients? Should It Be Drained or Stirred Back into the Yogurt?

That liquidy stuff is whey, the very stuff that Little Miss Muffet ate on a tuffet. When the bacteria that forms yogurt grows sufficiently, the milk coagulates. The proteins squeeze together and form curds, pushing out the watery whey.

Whey may be watery, but it isn't water. Whey contains sugar, minerals, some protein, and lactose. Don't waste it. Mix it back in with the rest of the yogurt. You'll be a better person (nutritionally, anyway) for it.

Submitted by Emanuel Kelmenson of Jericho, New York.

Why Do Most Yogurts Come with the Fruit on the Bottom? Why Not on the Top? Or Prestirred?

We had no idea that this topic so consumed yogurt lovers. But our dairy consultants indicated that yogurt lovers have strong feelings about where to put the fruit in their yogurt containers. Dairy expert Bruce Snow told us that there is no difference in the contents of fruit-on-the-bottom yogurts versus prestirred varieties, but added:

> Many people like the idea of stirring the cup and doing their own mixing. Some claim that they like to see just how much fruit is really in the cup! Other people couldn't be bothered and much prefer to have the premixed product.
>
> But if you notice, almost everyone stirs even the premixed yogurt to some extent. As the French say, *"Chacun à son goût!"*

Kent Sorrells, director of research and development at California dairy giant Altadena, marvels at the variety of preferences of yogurt eaters. Some prefer the prestirred varieties because they tend to have a slightly softer consistency. Others feel cheated if they can't stir the yogurt themselves. And some yogurtphiles, Sorrell notices, like to keep the fruit on the bottom, dipping in when desired. They alternate spoonfuls of pure yogurt with doses of fruit and yogurt, not unlike those of us who pick apart Oreos like biology students dissecting butterflies or who eat all the cake before devouring the frosting.

But not everyone eats yogurt straight out of the cup, Sorrell reminds us. Many folks like to eat yogurt out of a bowl. They tip the container upside-down into a bowl, and end up with what looks like a fruit sundae, with the topping where Nature intended it to be—on top.

Why don't yogurt makers try putting fruit on the top? The heavy fruit would sink anyway, and unevenly at that.

Submitted by Darcy Gordon of Los Altos, California.

Why Do You Need to Supply Oxygen to a Tropical Tank When Fish Are Quite Capable of Surviving Without Extra Oxygen in Lakes and Oceans? Why Do You See Oxygen Tanks More in Saltwater Aquariums Than Freshwater?

Robert Schmidt, of the North American Native Fishes Association, answers the first part of this Imponderable succinctly:

> You have to provide oxygen to any tank that has more fish (thus higher oxygen demand) than the plants and algae in that atmosphere can supply.

Most natural bodies of water are replete with oxygen-producing plants—by comparison, the plant life in a home tank is like a sprig of parsley next to a piece of halibut in a restaurant.

Oceans tend to have a richer and more abundant plant life than freshwater environments, but the absence of sufficient flora is not the only reason why saltwater aquaria require oxygen tanks, as Dr. Robert Rofen, of the Aquatic Research Institute, explains:

Salt water tanks need more aeration than fresh ... keep their inhabitants alive because the oxygen lev... water. With the added salt molecules present, there is les... the H_2O molecules for oxygen O_2 molecules to be present. ... absorbs less O_2 than does fresh water.

Submitted by Rudy, a caller on the Lee Fowler radio show, West Palm Beach, Florida.

What Happens to Criminals' Firearms Confiscated by Police During Arrests?

Somehow, we doubted that the police give criminals a claim check for their unlicensed Saturday night specials. The image of a hardened convict, just released from a federal penitentiary after ten years of hard time for armed robbery, being issued a new suit, twenty dollars, and his old, trusty semiautomatic didn't ring true.

So we conducted a survey of about ten police departments all over the United States to find out how the authorities contend with newfound criminal firearms. We discovered quickly that policies about confiscated guns are up to each jurisdiction and that their strategies vary wildly.

The first priority of all police departments is to hold firearms in case they will be needed as evidence in a trial. Corporal Joseph McQue, of the Philadelphia Police Department's Public Affairs Office, told *Imponderables* that guns are first taken to the ballistics unit, where they are checked for fireability. In order to prosecute someone under the Uniform Firearm Act, the weapon must be fireable. (Toting a nonfireable weapon may also be a lesser offense.)

Whether or not ballistics checks the weapon, most jurisdictions place guns in a property unit until the case comes to trial. In many areas, if the defendant is found guilty and the gun was taken to the courtroom as evidence, the firearm is taken back to the property unit til a judge rules upon its disposition. In most of the jurisdictions we urveyed, if the defendant is found guilty, the judge releases the gun to the police department.

And what do the police departments do with the weapons? Just about everything you can imagine:

1. Most commonly, police departments destroy confiscated weapons, and the preferred method seems to be melting. Firearms gathered by the Chicago, New York City, and Philadelphia police are usually melted and sold for scrap, although McQue added that in Philadelphia the melted metal is used to make manhole covers and sewer inlets. Pawnshop owners everywhere must heave a weary sigh when they ponder over the notion that guns worth five thousand dollars are melted alongside pieces of junk.

Some localities, such as Denver, prefer to crush guns and sell the flattened firearms for scrap. A few jurisdictions, such as Miami, sometimes merely disassemble firearms and trash them. This procedure is far superior to a method of disposal that Miami has long ago discarded: tossing unwanted guns in the ocean.

2. Some police departments use the guns that they confiscate. Most firearms confiscated from criminals are of low quality, but better pieces are often given to undercover police officers. In Louisiana, the Code of Criminal Procedure mandates that a court order will be issued to either destroy a weapon or to use it in the department. Police officer Christopher Landry told *Imponderables* how he obtained his duty gun:

> My duty weapon is now a Glock 17 (9mm), thanks to some juvenile delinquent unknown to me. Now that my department has approved .40 caliber for duty use, I'm going to call the judge again to put my name on the list [to obtain confiscated guns]. This judge has no problem with releasing weapons for duty use, but the full automatics get destroyed.

3. Some police departments sell confiscated guns, often at auctions, but the practice is becoming less common. According to several police officials we spoke to, selling firearms creates bad publicity; opens up potential liability problems; and perhaps worst of all puts more guns out on the streets.

An official at the Justice Department who wishes to remain anonymous told us that about ten years ago, when a certain jurisdiction decided to adopt a different model for its entire force, it sold all its guns back to the manufacturer.

And Officer Angelo Bitses, of the Miami Police Department Public Information Office, told us that on rare occasions, confiscated guns might be sold to another police department.

4. A few guns are donated to museums or are used for other police training.

5. John Kearney, a Chicago sculptor, provides perhaps the most uplifting use for confiscated guns. A friend of the artist was killed on his birthday by a neighbor who collected guns. Kearney was also deeply affected by the assassination of John F. Kennedy and Martin Luther King and joined the growing gun control movement. In the late 1960s, he created an outdoor sculpture, using seventy-five melted handguns, called "Hammer Your Swords into Plowshares."

This large piece is now owned by The Committee to End Handgun Abuse in Illinois. Every year, Kearney creates an original sculpture out of a gun used in a capital crime in the Chicago area, and the artwork is given to a politician or other notable who has worked to institute handgun control. Some of the recipients include: John Anderson, Sarah Brady, Pete McCloskey, and Dianne Feinstein.

Submitted by Christopher Stamler of Woodbridge, Virginia.

Why Is There a Worm on the Bottom of Some Tequila Bottles?

Because worms aren't good swimmers?

Those worms are a marketing concept designed to demonstrate that you've bought the real stuff. In order to research this topic with the rigorousness it deserves, we recently undertook a worm-hunting expedition to our local liquor store but found no tequila bottles with worms. We had heard about the worm-filled tequila bottles for years but had never found one ourselves.

So we beseeched one of our favorite liquor authorities, W. Ray Hyde, to help us. As usual, he knew the answer immediately.

We couldn't find worms in tequila bottles because they are included only in bottles of mescal, as he explains:

> Tequila and mescal are related beverages. Both are distinctive products of Mexico. While mescal is any distillate from the fermented juice of any variety of the plant *Agave Tequiliana Weber* (also known as

maquey), tequila is distilled from the fermented juice of only one variety of this plant and only in one restricted area of Mexico. Therefore, all tequila is [technically] mescal but not all mescal is tequila.

The worm is placed in bottles of mescal as an assurance that the beverage is genuine since the worm used lives only in the *Agave Tequiliana Weber* plant.

The worm is found only in the agave cactus in Oaxaca province. Natives of Oaxaca consider the worm a delicacy and believe that the agave possesses aphrodisiac powers.

Lynne Strang, of the Distilled Spirits Council of the United States, adds that the United States Food and Drug Administration approved the practice of allowing the worm in imported bottles of mescal and tequila in the late 1970s. Although actual worms were once the rule, most are now replicas, made of plastic or rubber.

Submitted by Suzanne Bustamante of Buena Park, California. Thanks also to Teresa Rais of Decatur, Georgia; Dianna Love of Seaside Park, New Jersey; Richard T. Rowe of Sparta, Wisconsin; Dana Patton of Olive Branch, Mississippi; Tim Langridge of Clinton Township, Michigan; Mary J. Davis of El Cajon, California; and Aaron Edelman of Jamesburg, New Jersey.

Do Fish Pee?

You don't see them swimming in your toilet, do you? Yes, of course, fish urinate.

But not all fish pee in the same way. Freshwater fish must rid themselves of the water that is constantly accumulating in their bodies through osmosis. According to Glenda Kelley, biologist for the International Game Fish Association, the kidneys of freshwater fish must produce copious amounts of dilute urine to prevent their tissues from becoming waterlogged.

Compared to their freshwater counterparts, marine fish, who *lose* water through osmosis, produce little urine. For those readers who have asked us if fish drink water, the surprising answer is that saltwater fish do, because they need to replenish the water lost through osmosis, as Kelley explains:

> This loss of water is compensated for largely by drinking large amounts of sea water, but the extra salt presents a problem. They rid themselves of this surplus by actively excreting salts, mainly through their gills.

Dr Robert R. Rofen, of the Aquatic Research Institute, told *Imponderables* that fish are able to excrete liquids through their gills and skin as well, "the counterpart to humans' sweating through their skin."

Submitted by Billie Faron of Genoa, Ohio.

Why Are Screen Door Handles and Knobs Located Higher than Their "Regular" Counterparts?

The height of door knobs and handles has become so standardized that we can usually find, say, a bathroom door knob in an unfamiliar hotel room, with the room pitch black. That's why it is so disconcerting to reach for the handle of a screen door and find that we are hitting the screen itself. Why aren't they the same height as "regular" hardware on doors?

According to Joe Lesniak, of the Door and Hardware Institute, screen doors are usually adjacent to another conventional (or sliding) door, which has its own fixtures. If the hardware on the screen door were at the same height as that on the conventional door, the fittings would conflict. This is also the reason why the door knobs in connecting hotel rooms are deliberately placed at "mismatching" heights.

So the height of screen (or storm) door handles is an afterthought. Most manufacturers choose to go higher with their screen-door fittings, but a minority go lower—anything to avoid a direct confrontation with the dreaded door knobs and handles of conventional doors.

Submitted by Lora B. Odom of Carmel, Indiana.

Why Aren't Automobiles Designed So That the Headlamps Shut Off Automatically When the Ignition Key Is Removed?

We venture to speculate that the three correspondents who submitted this question were motivated by a common experience: leaving the headlamps on inadvertently while running out of a just-parked car. Nothing makes one think more about headlamp design than having to buy a new battery.

Alas, the most common explanation for why headlamps stay on even when the ignition is shut off is that consumers want to be able to mark their vehicles' presence at night, particularly in emergency situations. Some drivers may also want to use headlamps, on occasion, as a giant flashlight, to illuminate a dark area in front of them. Still, we don't understand why these operation couldn't be performed by turning the auxiliary switch and turning the lights on manually. We came to the conclusion that an obvious reason why headlamps don't switch off automatically is simply because that's the way it has always been done.

Increasingly, automakers are listening to consumers' concerns on the matter. Richard Van Iderstine, of the National Highway Traffic Safety Administration, wrote *Imponderables* that there is no law governing the relationship between ignition and headlamp status but that many manufacturers are experimenting with delayed turn-off options for headlamps,

> allowing people to get out and see their way into their home at night. Others automatically turn the headlamp off . . . but leave the parking and tail lamps on to conserve the battery.

Vann H. Wilber, director of the safety and international department of the American Automobile Manufacturers Association, the organization that represents the Big Three Detroit automakers, told us that the current experiments with cutting headlamps automatically when the ignition is turned off are part of a long tradition of rolling out "limited introductions" of convenience features to selected models. If consumers prefer the change, automatic headlamp shutoff will join the illustrious list of now standard features that were once introduced only in limited introductions: automatic transmissions; power steering and brakes; radios; and air conditioning.

Submitted by Ed Leonardo of Arlington, Virginia. Thanks also to Jim Lyles II of Shreveport, Louisiana; Richard Tiede, Jr. of Mansfield, Georgia; and Christopher Doody of Shortsville, New York.

Why Do Many Blind People Wear Dark Glasses?

If David Letterman and Mr. Blackwell can do it, so can the Braille Institute: put out a top ten list, that is. Every year, the Braille Institute issues a list of the ten most unusual questions it receives. We are proud to report that this Imponderable made number nine on the 1993 list (edging out number ten: "Do blind babies smile?").

In our previous research on blindness, several authorities

emphasized that the majority of legally blind people do have some vision. The Braille Institute's answer to this question stresses the same point:

> Not everyone who is legally blind is totally blind. More than 75 percent of people who are legally blind have some residual vision. Blindness is the absence of sight, not necessarily the absence of light.

Alberta Orr, of the American Foundation for the Blind, adds, "Many visually impaired persons are extremely sensitive to bright light and glare and wear sunglasses to reduce the amount of light on the retina."

Some blind persons wear dark glasses for cosmetic purposes, because they are self-conscious about the physical appearance of their eyes. Increasingly, blind people are forgoing dark glasses, but we tend to associate dark glasses with blind people because so many of the high-visibility blind celebrities, such as Stevie Wonder, George Shearing, and Ray Charles, usually wear them. Even this is starting to change—the last time we saw José Feliciano perform on television, he was shadeless, despite the glare of the spotlights.

Submitted by Amy Kelly of Cleburne, Texas. Thanks also to Jim Wright, of New Orleans, Louisiana.

Why Do Many Fast Food Restaurants and Convenience Stores Have Vertical Rulers Alongside Their Main Entranceways?

On an episode of "The Simpsons," Marge was found shoplifting at the local convenience store. Her arrest was made considerably simpler when she passed the vertical measuring scale mounted along the exit doorway. It isn't often, even in a cartoon, that a suspect can be positively ID'd as an eight-foot woman, with a considerable percentage of that height consisting of bright blue hair.

Police officers we have spoken to over the years have regaled us with stories about how often witnesses supply them with unreliable descriptions of suspects. In particular, frightened witnesses tend to overestimate the height (and weight) of criminals. The ruler is an attempt to remedy flawed guesstimates.

We weren't able to locate any convenience store or fast food chain that installs measuring scales in all of its branches; obviously, scales tend to show up in urban, high-crime areas. Some chains, like Wendy's, never use scales. But Kim Bartley, director of marketing at White Castle, says that any White Castle store that wants one can install it; employees at those locations are instructed to view criminals leaving and observe their height as the miscreants take flight. Many White Castles, like convenience stores, are open at all hours, and more vulnerable to late-night stickups than their fast food competitors.

Submitted by Don Marti, Jr., of New York, New York. Thanks also to Viva Reinhardt and family of Sarasota, Florida.

What Does "100% Virgin Acrylic" Mean?

Acrylonitrile, the chemical substance from which acrylic fibers are derived, was first developed in Germany in 1893, but commercial production didn't begin until Du pont released Orlon in 1950. Monsanto, Dow Chemical, and American Cyanamid followed, all with their own trade names. Acrylic proved to be a durably popular wool substitute—it can be dyed more easily than wool, can be laundered easily, and is almost as versatile—and like wool, acrylics can be found in carpets as well as garments.

Our correspondent, Shirley Keller, was baffled by the meaning of the oft-found "virgin acrylic" label on many knit labels:

> Does this mean that the product comes from: a) the first polymerization of the Acryl; b) that the fiber was not previously woven; or c) is it a marketing scam to raise the price of the garment, a la "French" Dry Cleaning?

The answer is b. According to Bob Smith, of Cytec, the division of American Cyanamid that manufactures the product, "100% virgin acrylic" means that the material comes directly from the manufacturer and was never used before. Occasionally, acrylic fibers are reprocessed; just as with humans, acrylic fibers can be virgins only once.

The "100%" part of the label is a tad misleading. To be legally classified as acrylic, the fiber only has to be 85 percent acrylonitrile (by weight). According to Roscoe Wallace, chemical engineer for Monsanto, the other 15 percent may be comprised of other fibers, some of which may more easily allow dyeing or change the texture of the finished garment.

Actually, even our acrylic marketers were willing to concede that there is a bit of answer c in "100% virgin acrylic" labels. Sure, Monsanto's Larry Wallace was willing to concede, 100% virgin acrylic has no additives, it is not reworked after manufacturing, and was never reclaimed or redissolved. But even non-100% virgin acrylic must meet the same specifications as its more innocent brethren.

Submitted by Shirley Keller of Great Neck, New York.

Do Snakes Sneeze?

Norman J. Scott, Jr., zoologist and past-president of the Society for the Study of Amphibians and Reptiles, told *Imponderables*, "As far as I know, snakes don't sneeze with their mouths shut, but they do clear fluid from their throat with an explosive blast of air from the lungs."

Snakes don't sneeze very often, though. In fact, a few herpetologists we contacted denied that snakes sneeze at all. But John E. Simmons, of the American Society of Ichthyologists and Herpetologists Information Committee, insisted otherwise:

> Snakes sneeze for the same reason as other vertebrates—to clear their respiratory passages. Snakes rarely sneeze, however, and people who keep them in captivity know that sneezing in snakes is usually a sign of respiratory illness resulting in fluid in the air passage.

Submitted by Sue Scott of Baltimore, Maryland. Thanks also to June Puchy of Lyndhurst, Ohio.

Why Do Cookbooks Often Recommend Beating Egg Whites in a Copper Bowl?

We don't know whether any cookbook writers have received kickbacks from copper bowl manufacturers, but this advice always struck us as unnecessary and fussy. But then again, our cakes compare unfavorably to the offerings of school cafeterias.

We consulted our pals at the American Egg Board and United Egg Producers, and we learned there really *is* something to this copper bowl theory. The copper in the bowl reacts to a protein (the conalbumin, to be precise) in the egg whites, and helps stabilize the eggs and may actually increase their volume when whipped. Cream of tartar combines with egg whites in a similar fashion, working to keep the whites from separating from yolks. One reason why some cooks prefer to stabilize the whites with cream of tartar rather than the "no-cost" copper bowl is that if you leave the egg whites in the bowl for too long (sometimes, for as little as five minutes), the whites will turn pink.

Cooking is an art rather than a science, and we seem to see the prescription for the copper bowl less often these days. Kay Engelhardt, test kitchen supervisor for the American Egg Board, waxes philosophical:

> Perception of the copper bowl's merits varies considerably among various experts. The Strong Armed swear by it. The punier among us are willing to settle for an electric mixer and a bit of cream of tartar.

Submitted by Merilyn Trocino of Bellingham, Washington.

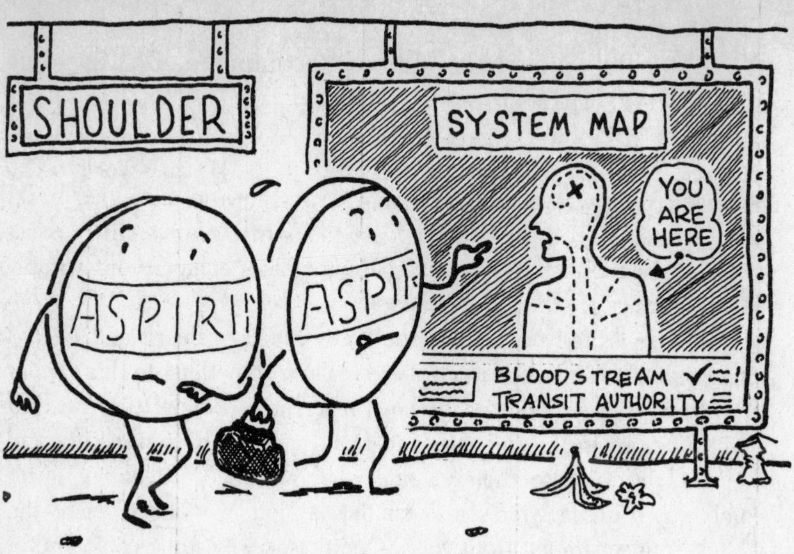

How Does Aspirin Find a Headache?

When we get a minor headache, we pop two aspirin and *voilà,* the pain diminishes within a matter of minutes. How did those little pills find exactly what ailed us instead of, say, our little right toe or our left hip?

We always assumed that the aspirin dissolved, entered our bloodstream, and quickly found its way to our brain. The chemicals then persuaded the brain to block out any feelings of pain in the body. Right? Wrong.

Willow bark, which provided the salicylic acid from which aspirin was originally synthesized, had been used as a pain remedy ever since the Greeks discovered its therapeutic power nearly 2,500 years ago. Bayer was the first company to market Aspirin commercially in 1899 ("Aspirin" was originally a trade name of Bayer's for the salicylic acid derivative, acetylsalicylic acid, or ASA). The value of this new drug was quickly apparent, but researchers had little idea how aspirin alleviated pain until the 1970s. In their fascinating book *The Aspirin Wars*, Charles C. Mann and Mark L. Plummer describe the basic dilemma:

> Aspirin was a hard problem. . . . It relieves pain but, mysteriously, is not an anesthetic. . . . And it soothes inflamed joints but leaves normal joints untouched. "How does aspirin "know" . . . whether pain is already present, or which joints are inflamed? Researchers didn't have a clue. They didn't even know whether aspirin acts peripherally, at the site of an injury, or centrally, blocking the ability of the brain and central nervous system to feel pain.

The breakthrough came more than seventy years after the introduction of the best-selling pharmaceutical in the world, when researcher John Vane discovered that aspirin inhibited the synthesis of prostaglandins, fatty acids manufactured by virtually every cell in the human body. They resemble hormones, insofar as they secrete into the bloodstream, but unlike most hormones, they tend to stay near their point of manufacture. Prostaglandins serve many biological functions, but the particular ones that cause headache pain, usually known as PGE 2, increase the sensitivity of pain receptors.

So the function of prostaglandins seems to be to produce discomfort, inflammation, fever, and irritation in areas of the body that are not functioning normally, thus serving as an internal warning system. According to Harold Davis, consumer safety officer with the Food and Drug Administration, prostaglandins dilate blood vessels, which can also produce headaches.

The discovery of the role of prostaglandins in producing pain explains why aspirin works only on malfunctioning cells and tissues; if aspirin can stop the production of prostaglandins, pain will not be felt in the first place. Still, aspirin doesn't cure diseases; it can alleviate the symptoms of arthritis, for example, but it doesn't stop the progress of the condition.

In all fairness, scientists still don't know exactly what causes headaches, nor all the ways in which aspirin works to relieve pain. Unlike morphine and other mind-altering drugs, aspirin works peripherally. The key to the success of any peripheral painkiller is in reaching the pain receptors near the irritation or inflammation, not simply in reaching sufficient concentrations in the bloodstream. In the case of aspirin, the ASA is connected to the bloodstream; the bloodstream's connected to the prostaglandins; the prostaglandins are connected to the receptors; and the receptors are connected to the headache.

Submitted by Debra Allen of Wichita Falls, Texas. We know we have received this question many times over the last five years, but we cannot find our records. If you previously submitted this question (postmarked before June 1, 1993), please let us know, and we'll include your name in future printings. Our apologies for this mistake.

Is Goofy Married? If Not, Where Did Television's Goofy, Jr., Come From?

Do you really expect Disney to give old Goof a child out of wedlock? We are pleased to announce that Goofy is, or possibly only was, married to a lovable mate named Mrs. Goofy.

Mrs. Goofy first appeared in a short, "Fathers Are People," but was far from salient; in fact, she can be seen only fleetingly. Although she was a Donna Reed-like suburban housewife, she had her husband well trained: Goofy's response to just about everything she ever said was, "Yes, dear."

Junior, with a red nose at the end of his snout and a mop of red hair on his head, was featured more often and prominently than his mother, but as Disney's Rose Motzko told *Imponderables*, "Goofy, Jr.'s main function was to allow his father to burst with pride while allowing his father not to live up to [minimal] expectations." Junior understood that his father was not a brain surgeon but tried hard not to let his father know.

On the current cartoon series *Goof Troop*, Junior is called "Max." Goofy is a single father, and Goofy's mother is never discussed. But come to think of it, most of the "family" TV sitcoms with live actors feature single parents, too: Max has plenty of company.

Submitted by Tai Palmgren of Davis, California.

Why Is "$" the Symbol for the American Dollar?

We remember reading a numismatics book thirty years ago that stated the $ was derived from a stylized version of an "S" superimposed on a "U." We never understood this explanation, because we could never see the "U" in the dollar sign. A professor of the history of mathematics at the University of California, Dr. Cajori, spent decades researching this Imponderable in the 1910s and 1920s. He concluded:

> The American dollar sign, popularly supposed to be derived from the letters U and S, is, instead a lineal descendant of the Spanish abbreviation "ps" for "pesos."

Cajori pored through hundreds of early colonial manuscripts and could find no proof of the "US" theory.

So, the official position of the Department of Treasury is that the "S" gradually came to be written over the "P" in the "pesos" abbreviation,

> developing a close equivalent of the $ mark, which eventually evolved [into our current mark]. The $ was widely used before the adoption of the United States dollar in 1785.

Indeed, as we discussed in *Why Do Clocks Run Clockwise?*, Spanish and Mexican coins were the main currency in many parts of the United States in the eighteenth and much of the nineteenth centuries. We're still not sure if the $ looks any more like a P and an S than a U and an S, but at least the abbreviation of "pesos" makes more sense than a shortening of "United States."

Submitted by Ed Booth of Chico, California. Thanks also to Ken Shafer of Traverse City, Michigan; Josh Siegel of Fountain Valley, California; and Barry Kaminsky of Brooklyn, New York.

What Do Paper Manufacturers Do with the "Holes" Punched Out of Looseleaf Paper? Do They Recycle Them?

You can bet your bippy that manufacturers recycle the liberated hole punches. But they differ in how they recycle. Forest products giant International Paper, for example, boils more than 90 percent of its paper byproducts to power the very plant that manufacturers looseleaf paper, according to IP representative Michael Goodwin.

But Mead's strategy is more common. Mary Potter, Mead consumer relations representative, told us:

> "Hole punches," as well as other types of paper trim or waste, are baled and sold for scrap. It is and has always been recycled (for approximately 100 years), usually winding up in chipboard, boxes, etc.

Indeed, when we asked this Imponderable of Fort Howard, its consumer affairs coordinator responded, "Nearly all Fort Howard products [mostly toilet paper, facial tissue, paper towels, and napkins] are made from 100% recycled paper fiber."

Bet you never considered that the missing dots in your notebook paper made their way into your toilet paper.

Submitted by Wendy Rath of Sandy, Utah. Thanks also to Alvin Polanco of Philadelphia, Pennsylvania.

Why Do We Bury the Dead with Heads Toward the West Facing East?

The following passage appears in Matthew 24: "For as the lightning cometh out of the east, and shineth even unto the west; so shall also the coming of the Son of man be."

Interpreting this as an indication that when Jesus is resurrected he will appear in the east, early Christians buried the deceased with the feet nearest the east and head towards the west (but facing east) so that the dead could best see and then hurry to rise up to meet Him. Dan Flory, president of the Cincinnati College of Mortuary Sciences, wrote *Imponderables* that this custom inspired the phrase "the wind of the dead man's feet" to describe an east wind.

The practice, both in Europe and North America, has steadily declined over time, but our informal observation is that the older the gravesite, the more likely the headstones will be situated in the western portion of the plot. In fact, burying the dead with this east-west orientation predates Christianity. Pagan societies, being sun worshipers, lay their deceased down to face the sunrise or sunset, depending upon the particular religion.

Submitted by Joseph Centko, Jr., of Streator, Illinois.

Why Do Birds Usually Take Flight Against the Wind?

Nancy Martin, naturalist at the Vermont Institute of Natural Science, points out that, given the constraints of runway design, airplane pilots prefer to take off against the wind as well. And for the same reason: It facilitates lift because of increased air speed. Martin elaborates:

> Birds' wings are structured like an airfoil and so work best with air flowing from front to back. Also, feathers are arranged to overlap like shingles to aid in smooth air flow—taking off with the wind ruffling up the feathers from behind creates a lot of useless turbulence.

Janet Hinshaw, librarian at the Wilson Ornithological Society at the University of Michigan, adds that birds with disproportionately heavy bodies for the size of their wings would probably take off against the wind more consistently—they can use all the lift they can get.

Submitted by Arpi Calioglu of Northridge, California.

Why Do Geese Honk Furiously While Migrating? Doesn't Honking Squander Their Energy on Long Flights?

Unlike humans, geese and other migrating birds don't have car radios and Stuckey's to keep them occupied on long trips. Honking allows geese to maintain communication during long flights. Most importantly, it helps them to avoid midair collisions. As Todd Culver, education specialist at the Cornell Laboratory of Ornithology, succinctly states, "Honking is cheap compared to crashing."

Culver adds that the call and response of birds is the main reason for flying in "V-formation." *Imponderables* has no desire to enter this raging debate, which we get asked about frequently. But our province is questions that aren't so well traveled. We have read theories about the etiology of V-formations ranging from greater aerodynamics to superior defense against predators and from facilitation of vision to Culver's theory about better auditory communication.

Janet Hinshaw, of the Wilson Ornithological Society, assures us that honking doesn't sap geese of vital energy: "They honk while exhaling, which they obviously have to do anyway."

Submitted by Steve Acheson of New Berlin, Wisconsin.

Why Do Scotsmen Wear Kilts? And Why Didn't Men in Surrounding Areas Wear Kilts?

Entire books have been written about the history of the kilt, so the first part of this question is hardly imponderable. Our reader's focus is on why this strange garment was a mainstay in the Highlands of Scotland and not in the rest of Scotland or surrounding countries.

Although we are most likely today to see a Scot in a kilt, inside or outside Scotland, only in a parade or on a formal occasion, its initial popularity was based on practical rather than ceremonial or aesthetic considerations. Although the contemporary kilt resembles a skirt, early kilts covered not only the waist to knee region of the body but the upper torso as well. Essentially, the earliest kilts were huge blankets, which were wrapped around the body several times and draped over the shoulder. This one garment served as blanket, sleeping bag, cloak, and trousers.

The geography of the Highlands of Scotland was no doubt responsible for the kilt's longevity. The Highlands are mountainous and damp, with innumerable streams and rivers. Anyone traversing the countryside in long pants and shoes would quickly be wearing wet long pants and wet shoes. The kilt saved the wearer from continually rolling up his pants. By rearranging the kilt, he could shield himself from the cold and wind. Perhaps most importantly, shepherds could leave their home base for months at a time wearing one garment and no "extra" clothes. As kilts were constructed out of elements easily obtainable in the Highlands (wool from the omnipresent sheep, and the plaid prints from native vegetable dyes), even the poorest of Highlanders could afford one. And the poor wore the kilt the most: According to Steward MacBreachan, a Scottish historian, performer, and demonstrator of Highland games and ancient Scottish culture, the kilt was of special importance to those who had to spend most or all of the day outdoors. More affluent Highlanders could switch from kilts to pants once they returned home from a day's work.

We had a long talk with Philip Smith, Ph.D., one of thirteen fellows of the Scottish Tartan Society worldwide and an author of several books about Scotland. He informed us that kilts, or their equivalents, were worn in many parts of Europe in the ancient world. The Scottish kilt is not too different from the garb of the ancient Romans and the Portuguese.

Smith feels that the widespread use of the horse in other countries eventually led to the abandonment of kiltlike clothing. For rather obvious anatomical reasons, kilts and horse riding are, let us say, an uncomfortable fit for men.

After an unsuccessful Jacobean uprising in 1745, the English Prohibition Act of 1746 (more commonly known as the "Dress Act") banned the wearing of both the kilt and any tartan material by anyone except the Highlands regiment. Ironically, the prohibition is probably responsible for our current association of Scotsmen with kilts. Scotsmen kept their kilts during the ban and wore them surreptitiously at closed gatherings. Along with the tartan, which identifies the clan of the wearer, the kilt became a symbol of Scottish pride.

As Scotsmen needed the blanketlike garment less and less for practical reasons in the nineteenth and twentieth centuries, the kilt, if anything, gained in significance as a way for Scotland to carve its psychic independence from England. If proof of this were necessary, we need only point to the wearing of kilts in ceremonial occasions by Scotsmen from the south, who never wore them in the eighteenth century.

Submitted by Yvonne Martino of La Verne, California.

Why Are the Muppets Left-Handed?

Our sharp-eyed correspondent, Jena Mori, first noticed that all the Muppet musicians seem to be left-handed, and then realized that just about all of the Muppets' complicated movements were done with their left hands. We went to the folks at Jim Henson Productions for the answer to Jena's conundrum and were lucky enough to get an expert answer right from the frog's mouth, so to speak.

Steve Whitmire has been a Muppet performer for fifteen years, and currently "is" Kermit The Frog. Steve performs Wembley Fraggle and Sprocket the Dog from "Fraggle Rock," as well as Rizzo the Rat, Bean Bunny, and numerous lesser-known Muppets. He also performs Robbie and B.P. Richfield on "Dinosaurs" and has worked on all of the Muppet movies.

Since we don't often have the opportunity to speak with Muppet performers, we imposed on Steve to answer in interview form.

> IMPONDERABLES: Steve, why are Muppets left-handed?
>
> STEVE: Because most puppeteers are right-handed.
>
> IMPONDERABLES: Huh?
>
> STEVE: Imagine standing with your right hand in the air. You are wearing a hand puppet that fits down to approximately your elbow. Now imagine that a television camera is raised to six feet off the floor and is pointing at everything above your head. You are watching what the camera sees on a television monitor on the floor in front of you. Your right hand is in the head of the character. If you want to move the puppet's arms, you reach up in front of your face and grasp one or both of the two wire rods that hang from the puppet's wrists. You have to make sure that your head is low enough to clear the camera frame, so you'll probably have to shift your weight to your left as you duck your head to the left.
>
> IMPONDERABLES: Why do you duck to your left instead of your right?
>
> STEVE: The right hand is stretching as high to the right as possible because that is most comfortable. When the right hand stretches up, the left side automatically hunches down a bit. It's easier for me to duck my head to the left; otherwise, I'd be ducking my head under my right arm.

IMPONDERABLES: If your right hand is controlling the head of the puppet, how are you controlling its arms?

STEVE: You reach up in front of your face and grasp one or both of the two wire rods that hang from the puppet's wrists. You'd be able to have general control of both arms with your left hand. If you needed to do some bit of action that is more specific, you'd likely use the puppet's left arm.

IMPONDERABLES: Aha, we're now at the crux of our Imponderable. But since you are controlling both of the puppet's arms with *your* left hand, why does it matter which of the *puppet*'s hands you control?

STEVE: Right-handed people tend to have more dexterity and stamina in their right hand and arm, so it goes into the head of the puppet. It is an ergonomic choice more than anything. If the puppeteer is right-handed, it is the more coordinated arm and hand, and it is usually best for it to be in the head. The left arm of the puppeteer is just below the puppet's left arm, so making the left hand of the puppet its dominant hand seems like the natural choice.

IMPONDERABLES: You are implying that a Muppet performer concentrates much more on the head of a character than its arms.

STEVE: The attention of the audience is generally focused on the puppet's face and, more specifically, its eyes. That's part of the appeal of the Muppets—they seem to be looking at whatever they are focused on, whether it is a prop, another character, or the home audience via the camera. The arms are somewhat secondary, although if they are performed badly, say, with arms dangling, they can attract unwanted attention.

Eye contact, and life within the face, is always the first priority in bringing our characters to life: simple head moves and gestures, accurate lip sync, etc., mimic human or animal movement. We keep all of the movement of the characters to the minimum needed to give them the life we want. There shouldn't be any movement without a purpose.

IMPONDERABLES: But some of the Muppets' movements seem awfully complicated. How can you control intricate movements with your "wrong" (i.e., left) hand manipulating two rods?

STEVE: If there is specific action that requires precision that would draw our attention away from the head for too long, we will often have another puppeteer handle the right, and occasionally both, hands.

IMPONDERABLES: Couldn't it get tricky having two people manipulate the same puppet?

STEVE: It can. Having one performer manipulating the head and left hand and another the right hand of the puppet can help. This method allows the puppeteer on the head to do any action with the left hand if it needs to come in contact with the face, or the puppet's right hand.

However, when Jim Henson did the Swedish Chef, he worked only the head, and it was usually Frank Oz in *both* hands. One reason for this was that the Chef's hands were actually human hands and needed to match. Another reason was that Jim and Frank loved to do difficult and silly things like that. Frank's goal was to break the china on the back wall each time they did a bit and the Chef threw something over his shoulder during his opening song. We would all take bets. I think he only did it [successfully] once or twice.

IMPONDERABLES: So this answers the question reader Robin R. Bolan asked about why some Muppets don't seem to have wires: The answer is that sometimes they don't.

STEVE: Right. These types of puppets are good for handling props because the puppeteer can simply pick things up. In this case, a second puppeteer *always* does the right hand of the character, because the lead performer is completely tied up with the head and left hand.

IMPONDERABLES: Sounds like it's easier to be green than a Muppet performer.

STEVE: I always liken what we do to being an air traffic controller, because there is so much to concentrate on while we are performing. Not only are we manipulating the puppet's mouth, body movements, and arms, we are doing the voice, remembering dialogue, watching a television screen (we never look at the puppet—only the screen), and tripping over cables, set pieces, and five other puppeteers who are doing the same thing we are.

It's a wonder we ever get anything done considering how truly complex it really is. Fortunately, and for good reason, the audience only sees what goes on up there above us.

Submitted by Jena Mori of Los Angeles, California. Thanks also to Robin R. Bolan of McLean, Virginia.

Why Do We Have a Delayed Reaction to Sunburn? Why Is Sunburn Often More Evident Twenty-four Hours After We've Been Out in the Sun?

It's happened to most of you. You leave the house for the beach. You forget the sunscreen. Oh well, you think, *I won't stay out in the sun too long.*

You *do* stay out in the sun too long, but you're surprised that you haven't burned too badly. Still, you feel a heaviness on your skin. That night, you start feeling a burning sensation.

The next morning, you wake up and go into the bathroom. You look in the mirror. George Hamilton is staring back at you. Don't you hate when that happens?

Despite our association of sunburn and tanning with fun in the sun, sunburn is, to quote U.S. Army dermatologist Col. John R. Cook, nothing more than "an injury to the skin caused by exposure to ultraviolet radiation." The sun's ultraviolet rays, ranging in length from 200 to 400 nanometers, invisible to the naked eye, are also responsible for skin cancer. Luckily for us, much of the damaging effects of the sun is filtered by our ozone layer.

Actually, some of us do redden quickly after exposure to the sun, but Samuel T. Selden, Chesapeake, Virginia, dermatologist, told us that this

> initial "blush" is primarily due to the heat, with blood going through the skin in an effort to radiate the heat to the outside, reducing the core temperature.

This initial reaction is not the burn itself. In most cases, the peak burn is reached fifteen to twenty-four hours after exposure. A whole series of events causes the erythema (reddening) of the skin, after a prolonged exposure to the sun:

1. In an attempt to repair damaged cells, vessels widen in order to rush blood to the surface of the skin. As biophysicist Joe Doyle puts it, "The redness we see is not actually the burn, but rather the blood that has come to repair the cells that have burned." This process, called vasodilation, is prompted by the release of one or more chemicals, such as kinins, setotonins, and histamines.
2. Capillaries break down and slowly leak blood.
3. Exposure to the sun stimulates the skin to manufacture more melanin, the pigment that makes us appear darker (darker-skinned people, in general, can better withstand exposure to the sun, and are more likely to tan than burn).
4. Prostaglandins, fatty acid compounds, are released after cells are damaged by the sun, and play some role in the delay of sunburns, but researchers don't know yet exactly how this works.

All four of these processes take time and explain the delayed appearance of sunburn. The rate at which an individual will tan is dependent upon the skin type (the amount of melanin already in the skin), the wavelength of the ultraviolet rays, the volume of time in the sun, and the time of day. (If you are tanning at any time other than office hours—9:00 A.M. to 5:00 P.M.—you are unlikely to burn.)

Even after erythema occurs, your body attempts to heal you. Peeling, for example, can be an important defense mechanism, as Dr. Selden explains:

> The peeling that takes place as the sunburn progresses is the skin's effort to thicken up in preparation for further sun exposure. The skin thickens and darkens with each sun exposure, but some individuals, lacking the ability to tan, suffer sunburns with each sun exposure.

One dermatologist, Joseph P. Bark, of Lexington, Kentucky, told us that the delayed burning effect is responsible for much of the severe skin damage he sees in his practice. Sunbathers think that if they haven't burned yet, they can continue sitting in the sun, but there is no way to gauge how much damage one has incurred simply by examining the color or extent of the erythema. To Bark, this is like saying there is no fire when we detect smoke, but no flames. Long before sunburns appear, a doctor can find cell damage by examining samples through a microscope.

Submitted by Launi Rountry of Brockton, Massachusetts.

Why Do Hockey Goalies Sometimes Bang Their Sticks on the Ice While the Puck Is on the Other End of the Rink?

No, they are not practicing how to bang on an opponent's head—the answer is far more benign.

In most sports, such as baseball, football, and basketball, play is stopped when substitutions are made. But ice hockey allows unlimited substitution *while the game is in progress*, one of the features that makes hockey such a fast-paced game.

It is the goalie's job to be a dispatcher, announcing to his teammates when traffic patterns are changing on the ice. For example, a minor penalty involves the offender serving two minutes in the penalty box. Some goalies bang the ice to signal to teammates that they are now at even strength.

But according to Herb Hammond, eastern regional scout for the New York Rangers, the banging is most commonly used by goalies whose teams are on a power play (a one-man advantage):

> It is his way of signaling to his teammates on the ice that the penalty is over and that they are no longer on the power play. Because the players are working hard and cannot see the scoreboard, the goalie is instructed by his coach to bang the stick on the ice to give them a signal they can hear.

Submitted by Daniell Bull of Alexandria, Virginia.

What Is the Substance That Resembles Red Paint Often Found on Circulated U.S. Coins? And Why Do Quarters Receive the Red Treatment More Often Than Other Coins?

The substance that resembles red paint probably *is* red paint. Or fingernail polish. Or red lacquer. Or the red dye from a marking pen.

Why is it there? According to Brenda F. Gatling, chief, executive secretariat of the United States Mint, the coins usually are deliberately "defaced" by interest groups for "special promotions, often to show the effect upon a local economy of a particular employer." Other times, political or special interest groups will mark coins to indicate their economic clout. Why quarters? As the largest and most valuable coin in heavy circulation, the marking is most visible and most likely to be noticed.

Some businesses—the most common culprits are bars and restaurants—mark quarters. Employees are then allowed to take "red quarters" out of the cash register and plunk them into jukeboxes. When the coins are emptied from the jukebox, the red quarters are retrieved, put back into the register, and the day's income reconciled.

Submitted by Bill O'Donnell of Eminence, Missouri. Thanks also to Thomas Frick of Los Angeles, California, and Michael Kinch of Corvallis, Oregon.

Why Are So Many Farm Plots Now Circular Instead of Squarish?

Our peripatetic correspondent Bonnie Gellas first noted this Imponderable while on frequent airplane trips. The neat checkerboard patterns of farm plots that she remembered from earlier days have transformed themselves into pie-plots. Are there hordes of agricultural exterior decorators convincing farmers that round is hip and rectangular is square?

We're afraid the answer is considerably more prosaic. The round farm plots are the result of modern irrigation technology—specifically, "center pivot irrigation" systems. Dale Vanderholm, associate dean for agricultural research at the University of Nebraska, Lincoln, told *Imponderables* that one of the problems with squarish plots was the expense required to water them. They required lateral movement systems, in which one huge pipe the length (or width) of the land traveled back and forth in order to irrigate the entire field.

Center pivot systems, on the other hand, require only one water source, at the center of each plot. Pipes must still move, but they travel only the relatively short distance around the "pivots." According to Lee Grant, of the University of Maryland's Agricultural Engineering Department:

> The traveling system moves on "tractors," spaced at intervals along the irrigation pipe. The "tractors" are supported by pairs of tractor type tires arranged one in front of the other. Motors driven by the flowing water turn the tires to pivot the irrigation pipe around the field.

We asked Vanderholm what farmers did with the "corners" of the circle, the small portions of land outside the reach of the spray. Usually farmers plant crops that don't require irrigation or don't farm that area. If they want to spend the money, they can also buy auxiliary arms, which can water areas beyond the reach of the center pivot.

According to Vanderholm, center pivot systems are most popular in the high plains states, just where you would likely be looking out the window on cross-country flights, bored out of your mind, craving sensory input, and seeking any alternative to the airplane food, movie, or seatmate.

Submitted by Bonnie Gellas of New York, New York. Thanks also to Gloria Klinesmith of Waukegan, Illinois.

Why Are Virginia, Massachusetts, Pennsylvania, and Kentucky Called "Commonwealths" Instead of "States"? What's the Difference Between a Commonwealth and a State?

These four states chose to call themselves commonwealths, yet trying to find a reason why they did so is a futile exercise. By all accounts, the word "state" preceded "commonwealth." Etymologists argue over whether the term predated medieval Europe, but all agree that the concept of "state" was well established by then. Most social scientists define "state" as any discrete political unit that has a fixed territory and a government with legal or political sovereignty over it. Theoretically, though, a "state," in its abstract form, could be taken over by a military dictator and retain its "stateness."

The notion of a commonwealth can be traced directly to the social philosophers of the seventeenth century, particularly Thomas Hobbes of England. He argued that the government should work for the "common weal" (or welfare) of the governed. These Hobbesian principles were articulated by John Winthrop, the first governor of the Massachusetts Bay colony, in 1637:

> the essential form of a common weale or body politic, such as this, is the consent of a certaine companie of people to cohabitate together under one government for their mutual safety and welfare.

All the historians we contacted thought that the four states called themselves "commonwealths" to emphasize their freedom from the monarchy in England and the republican nature of the government, while also indicating there was no evidence that they consciously tried to separate themselves from the other colonies that deemed themselves "states." Indeed, looking over the constitutions of the four commonwealths, we see that the crafters of the documents often used "commonwealth" and "state" interchangeably. For example, while the constitution of Virginia refers in several places to "the people of the Commonwealth" and "government for this Commonwealth," it also declares that "this State shall ever remain a member of the United States of America."

The Massachusetts Constitution, in the "Frame of Government" section, indicates its purpose:

> The people inhabiting the territory formerly called the province of Massachusetts Bay do hereby solemnly and mutually agree with each other to form themselves into a free, sovereign, and independent body-politic or State, by the name of the commonwealth of Massachusetts.

Of course, as far as the federal government is concerned, commonwealths are just like any other states, with all the privileges, rights, and taxes due thereto.

Submitted by Randall S. Varner of Mechanicsburg, Pennsylvania. Thanks also to Rick DeWitt of Erie, Pennsylvania, and William Lee of Melville, New York.

How Do Engineers Decide Where to Put Curves on Highways?

We don't expect to jolt any of you by announcing that the shortest distance between two points is a straight line. So one might think that it would be cheapest, most efficient, and most convenient to decide where a highway starts and where it ends, and then construct a straight roadway and link the two. But *one* would be wrong.

Sometimes it costs more to build in a straight line. For example, if a mountain happens to lie right in the middle of a proposed route, a consultation with engineers will reveal quickly that it is cheaper to direct the highway around the mountain than it is to level the natural formation.

If housing or commercial buildings are in the path of the "straight line," then it may be cheaper to lay extra cement and reduce the cost of buying out the more expensive land. Community opposition has killed more than one proposed highway, but evacuating dwellers and leveling buildings is not only a public relations disaster—it can be an economic one. As communities protest, delays create cost overruns.

Increasingly, environmental factors determine the curvature of a highway. Joan C. Peyrebrune, technical projects manager for the Institute of Transportation Engineers, told *Imponderables* that while highway officials have always been concerned about the impact of new highways upon existing houses and businesses, they are now equally aware of how highways might affect wetlands, parklands, wildlife habitats, etc.

If topography, economics, or community pressure necessitates curves in the highway, engineers are first concerned about safety. Peyrebrune notes that many studies have been conducted about the relationship between the radius of curvature and what speeds can be navigated safely on roadways. The American Association of State Highway and Traffic Officials publishes tables that determine the proper speed limits for given radii of curvature and slopes of highways.

Thomas Deen, executive director of the Transportation Research Board of the National Research Council, provided us with a "philosophy of curvature" that echoed our other sources:

> The alignment should consist of long, gentle curves with straight tangent sections for passing on two-lane roads. The design speed of the highway is the limiting factor on the minimum radius of curvature, but the alignment should if at all possible incorporate flat, long curves with the smallest central angle. The alignment should be consistent and not incorporate short, sharp curves with long, straight tangent sections. The choice of highway alignment is made to facilitate the motoring public's trip from one location to another in a safe, efficient, and pleasing manner.

Properly designed curvature doesn't render a highway less safe than a straight roadway. In fact, Peyrebrune indicates that "curvature of highways is preferred to long, straight stretches by most motorists," who can be lulled into obliviousness by the dullness of a straight route.

We have learned quite a bit about how to construct highways in the last few hundred years—it wasn't always a science. Thomas Werner, director of the traffic engineering and safety division of the New York State Department of Transportation, explained to us how one city wound up with its "distinctive" asymmetrical street structure:

> In colonial Boston, cattle once roamed the city. Where they trampled the grass and wore a path, colonialists used the trail to transport goods from one point to another. These cow paths soon became the basis of the city's street network.

Evidently, colonial cows didn't walk in a straight line, either.

Submitted by Rory Sellers of Carmel, California.

The great prima ballerina Bragonova entertains her biggest fans...

Why Are There So Many Different Types of Wine Glasses? Would Champagne Really Taste Worse If Drunk Out of a Burgundy Glass?

We have always been a tad suspicious about the pretensions of wine connoisseurs, and we, too, have wondered whether the "flute" glass for champagne truly enhances the taste of the bubbly wine. We were shocked when we found out that Riedel, the Austrian specialist in glassware for wines, now sells twenty-three different glass types—each is designed to be used with one particular variety of wine.

To help answer this Imponderable, we contacted Pat McKelvey, librarian of the Wine Institute in San Francisco, California. She told us that appropriate glassware meets three criteria:

1) It is thin and clear, to best show off the beauty of the wine.

2) Its shape is best suited to enhance and accentuate the natural bouquet of the wine.

3) Perhaps most importantly, the shape of the rim should direct the wine onto the appropriate portion of the tongue.

Our tongues are full of taste buds—four distinctive types. The buds at the tip of the tongue are most sensitive to sweetness; the buds at the edges are most sensitive to salt (which is why we put salt on the edges of tequila glasses); the buds at the sides of the tongue are most sensitive to acidity; and the buds at the back of the tongue are most sensitive to bitterness. Until recently, most glassmaking technology has focused on designing the appropriate shape and size of the bowl of the glass. The deep, narrow champagne flute was designed to conserve and accentuate the bubbles; the wide burgundy glass, tapered at the top, attempted to catch and release the fruity aroma, while letting in as little ambient air as possible that might dissipate the wine's character.

But even if the bouquet were enhanced by the shape of the glass, it meant little if the wine didn't *taste* better; this is why the emphasis, increasingly, is on "rim technology." In a young burgundy, for example, the high acid level can sometimes overcome the desired fruity taste. The solution, in this case, was to flare the rim so that the wine hit the tip of the tongue, which detects the sweetness of the grapes.

Some wines tend to become unbalanced, with the acid/fruit quotient at one extreme or the other. This is one reason why wine lovers swirl the filled glass. A cabernet Sauvignon glass is wide, so that swirling will blend the flavors more easily. The mouth of the glass is narrow, so that when you drink from it, the liquid hits the middle of the tongue. The proper cabernet Sauvignon glass is designed to hit all four types of taste buds each time you take a swallow.

Riedel has changed the shape of the classic German Riesling glasses. Riesling used to be sweeter, so glasses were designed to direct the wine to the sides of the mouth, where the buds would detect acids more acutely. But now that vintners have made German Rieslings more dry by introducing more acids into the wine, Riedel's glass has an outturned rim, in order to guide the wine directly to the tip of the tongue, where the wine's sweetness will be perceived first.

Riedel tests the efficacy of its designs by conducting blind taste testings, in which the same wine is poured into many different glass configurations. If the "right" glass is not preferred, then Riedel knows it's time to go back to the drawing boards.

Of course, if your preference is for wine that comes in screw-top bottles, disregard the foregoing.

Submitted by Adrienne Ting of Laguna Hills, California.

Why Do Soft Breads Get Hard When They Get Stale While Hard Starches Like Crackers Get Softer When Stale?

Staling bread is a perfect example of reversion to the mean. Bakery consultant Simon Jackel told *Imponderables* that the typical soft bread contains 32 to 38 percent moisture. If the bread is left unwrapped and exposed to the elements, it will become hard when it lessens to about 14 percent moisture.

Why does the bread get stale and lose the moisture? Although food technologists don't fully understand all the causes, a process called "retrogradation" occurs, in which internal changes take place in the starch structure. Although bread items are formulated to have a softer crumb portion than crust area, during retrogradation some of the crumb moisture migrates to the crust, which results in the softening of the crust and a hardening of the crumb.

Tom Lehmann, of the American Institute of Baking, adds that as the bread retrogrades, "a portion of the starch in the flour undergoes a gradual change, known as "crystallization," which results in a gradual firming of the bread. Some of the edible ingredients in the dough, such as enzymes and monoglycerides, act to slow up the rate of retrogradation, but the process is inevitable and will occur quickly if the bread is unwrapped and exposed to air.

Hard starches, such as crackers, are crisp because they are baked with an extremely low moisture level, usually 2 to 5 percent. When they soften, their internal structure doesn't change like staling hard breads. As they are exposed to the ambient air, crackers absorb the air's moisture. According to Jackel, hard crackers will be perceived as soft once the moisture level reaches 9 percent.

Submitted by Robert Prots of Waverly, Ohio. Thanks also to Jim Clair of Philadelphia, Pennsylvania.

Why do Hardcover Books Have Exposed, Checkered Cloth as Part of Their Bindings (on Top and Bottom)?

We are ashamed, almost morose, about the fact that we have spent nearly forty years of our lives reading books and never noticed the checkered cloths until reader Valerie Y. Grollman came into our lives. After we received Ms. Grollman's letter, we fondled many books in our collection and discovered that just about all our hardbound books did, indeed, have this embellishment on top and bottom.

We contacted several publishing authorities and received a fascinating historical explanation from Gerald W. Lange, a master printer associated with Los Angeles's Bieler Press. His remarks were so interesting that, with his permission, we are quoting them at length:

> The "cloth" you refer to is the "headband," which was originally a cord or cloth tape with colored thread or string tightly wound around it. The headband was an integral part of the binding structure in early forms of the "codex" book. The codex has been with us for nearly two thousand years and is the physical form of the book we are all familiar with today. (The primary ancestral form of the book prior to the codex was the papyrus roll.)
>
> In early examples of the codex, the thread winding around the cord actually pierced through into the folded "signatures" [grouping of pages] of the book's "text block" [the finished, sewn gathering of the signatures] and the cord itself was tied to the edges of the binding's casing at the head (upper) and tail (lower) of the spine, and as such, provided a great deal of structural strength to the binding.

> Some historians have suggested that a later and more cosmetic function of the headband was to hide from view the internal casing material of the binding's spine. Early Western books were part and parcel of a pervasive religious world view and any visually displeasing structural imperfections were hidden or disguised with decoration. Long after original intent, of course, traditional practices remain in place.
>
> With the development of commercial bookbinding production during the Industrial Revolution, the headband became less structurally important and was merely attached to the edges of the text block, rather than sewn into it. Today the headband serves a purely decorative purpose, and is now more often a thin strip of colored or patterned cloth glued to the edge of the spine. There are still a few fine craft hand binders who will take the time to provide headbands in the "old fashioned" way.

Lange mentions that a second theory has been advanced to explain the origins of the headband: It might have served as a buffer to support the spine material "at its vulnerable edges." We tend to pull books from a shelf by yanking on the top of the spine and pulling the spine backwards, which places obvious stress upon the cover. Lange dismisses this theory as the original reason for the headband, since the codex book, for more than one thousand years of its existence, was designed to lie flat, not upright. Bookshelves didn't exist, and most books were too heavy and cumbersome to stand upright.

Stephen P. Snyder, executive vice-president of the Book Manufacturers Institute, concurs with Lange's historical assessment and adds that headbands have sometimes served the purpose of hiding glue that has seeped out of the adhesive bindings of books. When the headband is applied mechanically, as they are in most commercial books today, the bands are fed from a big spool. With the spiraling cost of hardcovers, it is nice to know that one step on the assembly line is there merely to apply some thread to make your book a little prettier.

Submitted by Valerie Y. Grollman of North Brunswick, New Jersey.

Will Super Glue Stick to Teflon?

We were wary of contacting Loctite and Teflon about this almost metaphysical Imponderable, for it would be like prying a confession from the immovable object (Teflon) and the unstoppable force (Super Glue) that one of their reputations was seriously exaggerated. But we are worldly wise in such matters. After all, we had already cracked the centuries-old conundrum about "If nothing sticks to Teflon, how do they get Teflon to stick to the pan?" in *Why Do Clocks Run Clockwise?* We were ready for a new challenge.

So first we contacted Du Pont, the chemical giant that markets Teflon, a registered trademark for polytetrafluoroethylene (which, for obvious reasons, we'll call ptfe). As we expected, Kenneth Leavell, research supervisor for Du Pont's Teflon/Silverstone division, took a hard line. He firmly holds the conviction that Super Glue won't stick to Teflon, at least "not very well and certainly not reliably." Here are some of the reasons why not:

> 1. The combination of fluorine and carbon in ptfe forms one of the strongest bonds in the chemical world and one of the most stable.
>
> 2. The fluorine atoms around the carbon-fluorine bond are inert, so they form an "impenetrable shield" around the chain of carbon atoms, keeping other chemicals from entering. As Leavell puts it,
>
> > Adhesives need to chemically or physically bond to the substrate to which they are applied. Ptfe contains no chemical sites for other substances to bond with.
>
> 3. As we just learned with glue bottles, adhesives need to wet the substrate directly or creep into porous areas in the substrate. But the low surface energy of ptfe prevents wetting and bonding. Leavell compares it to trying to get oil and water to stick together.

And then he lays down the gauntlet:

> Super Glue is "super" because of its speed of cure and relatively strong bonds. As an adhesive for ptfe, it's no better than epoxies, polyurethanes, etc., would be.

So, the immovable object claims near invincibility. How would the unstoppable force react? We contacted Loctite's Richard Palin, technical service adviser. And he folded like a newly cleaned shirt. Yes, Palin admitted, Teflon lacks the cracks necessary for Super Glue to enter in order to bond properly; there would be nowhere for the glue to get into the pan. Yes, he confessed, the critical surface tension is too low for the adhesive to wet the surface. Yes, he broke down in sobs, Super Glue would probably just bead up if applied to a Teflon pan.

Just kidding, actually. Palin didn't seem upset at all about Super Glue's inability to stick to Teflon. By all accounts, there doesn't seem to be much demand for the task.

Submitted by Bill O'Donnell of Eminence, Missouri.

Why Do Many Women's Fingernails Turn Yellow After Repeated Use of Nail Polish?

Chances are, the culprit is one of two types of ingredients contained in all nail polish:

1. *Nitrous cellulose.* Nitrous cellulose is wood pulp treated with acids; it provides the hardness necessary to make polish stay on your nail plate. As a senior chemist from a major cosmetics company told us: When the moisture from polish dries, all you are left with is nitrous cellulose and pigment.

When nitrous cellulose breaks down, nitric acid forms. Nitric acid attacks the proteins in the nail and turns the nail yellow. The yellowing occurs only on the top layers of the nail plate and will eventually fade away if more acids aren't applied.

2. *Preservatives.* Tolulene-sulfonimide and formaldehyde are part of the base coat of nail polishes and are used as preservatives. According to dermatologist Jerome Litt, of Beachwood, Ohio, formaldehyde resin can turn keratin (the tough, fibrous protein that is the principal constituent in nails) yellow.

But don't assume that yellow nails are necessarily caused by the chemicals in nail-care products (or other, noncosmetic chemicals, such as inks, shoe polishes, and dyes, which can also stain nails yellow). Many heavy smokers, for example, have yellow nails. Even the ingredients in some orally administered pharmaceuticals can stain nails.

Physicians often examine fingernails to help determine the general health of patients, for many illnesses are betrayed by yellowing. The most common noncosmetics cause of yellowed nails is a yeast or other type of fungus infection underneath the nail plate. But many other, more serious illnesses can occasionally be diagnosed when a physician spots yellow nails, including diseases of the lymphatic system, thyroid, chronic respiratory disease, diabetes, and certain liver and kidney diseases.

Other maladies are tipped off by different-colored nails. If the normal Caucasian nail looks pink, because of the ample blood supply to the nail bed underneath the nail plate, it can turn white when a person is anemic, and blue if the patient is suffering from heart or lung disease and insufficient oxygen is sent to the nail bed.

Now that we've scared you sufficiently, we'll remind you that nail polish is much more likely to cause the discoloration than the illnesses we've chronicled above.

Submitted by Barbara Forsberg of Ballston Spa, New York.

Why Are Most Corrugated Boxes from Japan Yellow?

Unless it is bleached, the color of a box will be the color of its main material source—the wood that is turned into pulp fiber. In Japan (and China), the most plentiful source of fiber is straw, which has a yellow color, whereas North America's main source for pulp is tannish trees.

Robert H. Gray, vice-president of the corrugated division of Old Dominion Box Company, told *Imponderables* that demand for paper is so great that most countries turn to local fiber sources that can be "easily grown and harvested in volume." Corrugated boxes often contain recycled fibers besides wood, and Japanese boxes tend to have a higher percentage of recycled material in their pulp, both for ecological reasons and because straw fibers are weaker than wood fibers and bond less effectively.

As a result, the natural shade of Japanese boxes is more variable than our reliably colored kraft tan boxes. We don't know whether the yellowish tinge of straw is what motivated Japanese boxmakers to dye their boxes yellow, but that is indeed what they do, as James F. Nolan, vice-president of the Fibre Box Association, explains:

> The paper used in Asia for corrugated boxes is primarily recycled—with highly mixed sources of waste paper. In order to provide a uniform color for good print quality, the paper for the outer sheet of the corrugated board must be dyed.

Not that American boxes are beyond dye jobs. Although tan boxes are not dyed, liner-board white boxes are, according to Jim Boldt, of corrugated container giant Great Northern.

One glimpse of what undyed paper might look like was supplied by Karl Torjussen, of Westvaco. He asked us if we could think of the color of the cardboard backing on legal pads. "Sure," we responded, "sort of a dishwater gray."

That gray is the *natural,* undyed color of newsprint that has *not* been de-inked. No one cares too much what the back of a legal pad looks like, but we might find a stack of gray boxes utterly depressing—which is why, if boxes are made out of mostly recycled material, we dye them white and Asians dye them yellow.

Submitted by Kirk Baird of Noblesville, Indiana.

Welcome to
MODESTY, Vt.
"Even our bridges
are covered"

Why Are Covered Bridges Covered?

We have driven by stretches of rivers where, it seemed, about every third bridge we passed was a covered bridge. Why is one covered when the next two are topless?

The most obvious advantage to a covered bridge is that it blocks "the elements," particularly snow. Accumulated snow can render a bridge impassable, and it is true that covered bridges are found most often in cold climates. Of course, one could argue that engineers should design covers for all roadways. But as we learned in *Do Penguins Have Knees?* (ah, but have we retained it?), bridges remain frozen long after adjacent road surfaces, primarily because bridge surfaces are exposed to the elements from all sides, the bottom as well as the top.

But then some folks believe that covered wooden bridges were originally constructed to ease the fears of horses, who were skittish about crossing bridges, particularly if they saw torrents of water gushing below. The fact that covered bridges resembled wooden barns supposedly also allayed the horses' anxiety.

This question is reminiscent of one of our chestnut-Imponderables: Why do ranchers hang boots upside-down on fenceposts? The most likely answer is the same for both: to save wood from rotting. Alternate cycles of rain and sun play havoc on the wood. According to Stanley Gordon, of the Federal Highway Administration's Bridge Division, an uncovered wooden bridge might last twenty years, while a covered bridge can last a century or longer.

Submitted by Gary L. Horn of Sacramento, California. Thanks also to Matthew Huang of Rancho Palos Verdes, California.

How Do Waiters and Waitresses Get Their Tip Money When the Gratuity Is Placed on a Credit Card?

Let's look at the life cycle of a credit card transaction at a restaurant:

1. You get the bill. You pull out your trusty credit card and hand it to the waiter.
2. In most restaurants, the waiter or other employee must authorize the charge. In all but small restaurants, this authorization is now done electronically, and increasingly, a "record of charge" is printed out automatically displaying the food, liquor, and tax charges. Some restaurants still use the old-fashioned "chit," in which the amounts must be entered by pen.
3. Conveniently (for the wait staff, anyway), a gaping space is left for "gratuity," and the total amount of the bill is left empty.
4. The waiter also puts his name or ID code on the record of charge (this is the strange, unidentifiable number often put in a box toward the bottom of credit card chits).
5. If the charge has been approved, the waiter brings the paperwork back to the customer.

This is the point at which many diners have no idea what to do. For some reason, the myth persists that waiters do not want you to charge their gratuity. Walter Sanders, director of corporate affairs for

Citicorp Diners Club (Citicorp bought Diners Club in 1980), phrased the dilemma so charmingly that we are allowing him a blatant plug:

> My dad is one of those people so concerned about waiters and waitresses getting their tips that even when he charges a meal (on the Diners Club Card, of course) he still painstakingly digs for a *cash* tip, which he leaves under the coffee saucer.
>
> Well, your readers—and my dad—can now rest assured that waitpeople everywhere get their full tips, in cash, even when those tips are put on the Diners Club Card.

6. Let's assume the customer does pay the gratuity using a credit card. When the restaurant closes, the night's proceeds are tabulated, and the waiter is paid on the spot, in cash (some restaurants pay on a weekly basis). The ID number of the waiter is used to identify his share of the gratuity money (although some restaurants pool tips). According to Melissa A. Bertelsen, of First Data Resources,

> most merchants have Electronic Data Capture devices that will allow the merchant to enter the amount of the ticket and tip. The device will then break out the tip by waiter number and total his amount in tips for the evening.

7. Most credit card companies pay the merchant's bank, electronically, the total amount charged, often within twenty-four hours. This quick transfer of money is one of the reasons why restaurateurs are willing to pay the "discount fee" that allows the credit card companies to make money. While it may take six weeks for the credit card company to be paid for the restaurant bill by the consumer, the restaurant is paid within a day or two. Thus the credit card issuer has to "eat" the float—the "free" use of the value of the charge—that the consumer has been granted.

The negative float of credit card issuers, especially those, like American Express, that do not impose (high) finance charges for late payments by consumers, is one of their major costs of business. Nobody, especially you, is going to get rich by not having to pay a bill of fifty dollars for forty-five days, but imagine the impact of a financial institution contending with twenty million consumers receiving an interest-free loan for that period. Premium cards, such as American Express, try to regain the revenue lost in finance charges by pricing their cards higher than Visa or Mastercard, and by charging merchants a higher-percentage fee.

If there is any reason why waiters might prefer cash, it might have to do with three letters—IRS. Obviously, tips put on credit cards leave a paper trail; increasingly, the IRS is trying to find ways to chase after undeclared income from workers who derive most of their money from tips.

Submitted by Maria Scott of Cincinnati, Ohio.

Why Do Beacons on Police Cars Flash Blue and Red Lights? Why Are the Blue Lights on the Passenger's Side and the Red Lights on the Driver's Side?

Police work is serious business. We've always wondered why officers in fast pursuit of bad guys flash two-tone signals from their beacons. We called many big-city police departments to find out who their color consultant was.

We quickly discovered that there is no national law specifying the colors on police car beacons. Yet in practice, the choices are few. A law enforcement information specialist at the National Criminal Justice Reference Library who wished to remain anonymous told *Imponderables* that at the time when red and blue lights were chosen for most police department beacons, high-intensity lights were not in use. So there was a practical advantage to using two colors—blue was easier to see during the day, and red was more clearly discerned at night.

From time to time, there have been attempts to make yellow (the easiest color to observe from long distances) the official color of beacons throughout the United States, but the expense and effort of defying tradition and passing the legislation have killed such attempts. A federal regulation would cause disruptions in states like Pennsylvania, which have laws designating the color of beacons (in this case, blue and red for police). And opponents argue that yellow flashing lights would be confused with construction or street lights.

Blue was probably chosen initially for its long association with police (e.g., blue lights in front of police stations, blue uniforms), and because of its high daylight visibility. And red has long been a symbol of warning and danger, and a signal to stop. Police departments in Los Angeles, Dallas, Detroit, Philadelphia, and Salt Lake City all use the red and blue beacons. Chicago and the Virginia State Police, on the other hand, have switched from red and blue to all-blue beacons.

Several of the police officers we contacted argued that blue is the most effective color for beacons because no other emergency service uses it (both firefighters and ambulances use red beacons, and most construction and emergency transport cars employ yellow or amber). According to Bill Dwyer, of beacon manufacturer Federal Signal Corporation, big-city police departments, in particular, tend to prefer blue beacons, because the color distinguishes them from the many other emergency vehicles. And no other emergency vehicle features a two-colored beacon.

Why is the red light on the driver's side? We received the same answer from everyone, but Officer Romero, of the Los Angeles Police Department, put it best:

> The reason that the red light is over the driver's seat is so that the driver being pursued can better see it. People are conditioned to stop for a red light; this is the most efficient way to signal the driver of a car in front of you to stop.

A passenger in the offending car cannot see the red nearly as well as the blue light. The LAPD uses an amber light on the rear of the car, which is activated by an on-off switch. We are also conditioned to think of a yellow light as a caution light; in this case, cars behind the police vehicle are being cautioned by the amber lights to slow down because police activity is taking place.

Police departments are constantly experimenting with color possibilities. The Virginia State Police experimented with blue lights for four years with equipment from six manufacturers before adopting them. Maryland tried a multicolor approach: Different colors hooked to switches controlled inside the car, with the intention of color-coding specific activities. In hot pursuit of a car, all-red might be appropriate;

for a routine traffic ticket, red and blue might do the trick. And yellow would be the perfect understated fashion statement for lurking around a bend on a highway speed trap. But the state decided such color tactics were altogether too subtle and abandoned the idea.

Submitted by Ronald Lindow of Pittsburgh, Pennsylvania. Thanks also to Jim Wright of New Orleans, Louisiana, and Sean O'Melveny of Littleton, Colorado.

Why Is It So Hard to Find Single-Serving Cartons of Skim or Lowfat Milk? Why Is It So Hard to Find Single-Serving Cartons of Whole-Milk Chocolate Milk?

Obviously, milk distributors *do* make single-serving sizes of lowfat milk and skim milk. They can be found in schools and institutions throughout the country. As Paul E. Hand, secretary and general manager of the Atlantic Dairy Cooperative reminded us, single-serving cartons of lowfat milk are a staple at McDonald's, Burger King, and many other fast food establishments. Hand added that many school lunch programs do include prepackaged chocolate whole milk in one-cup cartons.

So why can't you find them in the supermarket? In some cases, you can. But grocery stores want to stock a limited number of container sizes and prefer selling big containers to small ones to maximize profits. (Note the demise of the seven-ounce soda bottle while three-liter containers proliferate on supermarket shelves.)

Dairy distributors realize economies of scale by saving on packaging costs (obviously, four single-serving packages of milk are required to provide the milk in one quart container). Milk is a staple in most households, one used on a daily basis; Hand reports that there simply isn't sufficient demand for single-serving cartons. If consumers bought them or demanded them in sufficient quantity, they'd be on the shelves.

Prepackaged chocolate milk, an insignificant category two decades ago, while steadily gaining in market share, is still a stepchild to unflavored milks. In most cases, supermarkets don't want to stock more than one type of chocolate milk. For example, Hershey, the closest to a national brand in this category, licenses local dairies to produce its brand of chocolate milk. Individual dairies can choose whether to use 2 percent or whole milk. According to Hershey Foods' Carl Andrews, most local dairies will base their decision on whether to use whole or lowfat milk by assessing which type of unflavored milk sells better in their region.

You live in southern California, Mitch, where 2 percent milk dwarfs the sale of whole milk. In many parts of the East coast of the United States, you'd have a hard time finding single-serving cartons of lowfat chocolate milk.

Submitted by Mitch Hubbard of Rancho Palos Verdes, California.

What Happens to the Ink When Newspapers Are Recycled?

Before used newsprint can be recycled, it must be cleaned of contaminants, and ink is the most plentiful contaminant. The newsprint must be de-inked.

Although synthetic inks are gaining market share, most newspapers still use oil-based inks. To clean the newspaper, the newsprint is chopped up and boiled in water with some additional chemicals until it turns into a slurry. As the fibers rub against each other, the ink rises to the surface, along with other nuisances, such as paper clips and staples. A slightly different, more complicated procedure is used to clean most newsprint with polymer-based inks.

Theodore Lustig, a professor at West Virginia University's Perley Isaac Reed School of Journalism, and printing ink columnist for *Graphic Arts Monthly*, stresses that current technology is far from perfect:

> You should be aware that it is impossible to remove *all* ink from the slurry prior to recycling it into new paper. Since microparticles of ink remain, this would leave the paper rather gray if used without further processing. It is often subjected to bleaching or is mixed with virgin fibers to increase the finished recycled paper's overall brightness, a requisite for readability contrast.

More and more states are requiring publishers to use a higher proportion of recycled paper. As recyclers extract more ink from more newsprint, it may save trees in the forest, but it results in another ecological problem: what to do with unwanted ink. Although we may think of ink as a benign substance, the EPA thinks otherwise, as Lustig explains:

> The ink residue is collected and concentrated (i.e., the water is removed) into a sludge for disposition. However, since there are trace elements of heavy metals (lead, cadmium, chrome, arsenic, etc.) in this residue, this sludge is considered by EPA and other agencies to be a hazardous waste and has to be disposed of in accordance with current environmental laws.

In the past, sludge was dumped in landfills. Today, many options are exercised. According to Tonda F. Rush, president and CEO of the National Newspaper Association, some mills burn the waste, while others sell it to be converted to organic fertilizer.

Recycled newsprint can feel differently to the touch than virgin stock. Lustig explains that paper cannot be recycled infinitely. Three or four times is a maximum:

> Eventually, the fibers lose their ability to bind together, resulting in a paper that is structurally weak and unable to withstand the tensile pressures put to it on high-speed web presses.

Submitted by Ted Winston of Burbank, California. Thanks also to Meadow D'Arcy of Oakland, California.

Why Do Lizards Sleep with One Eye Open?

We imagine that Sarah Robertson, who lives in Nevada, has had ample opportunity to observe lizards sleeping, but our experts beg to differ with this Imponderable's premise. Professor Joseph C. Mitchell, secretary of the Herpetologists League, says that it is rare for lizards to sleep with one eye closed. Norman J. Scott, Jr., zoologist in lizard country (New Mexico), says, "Every lizard that I caught sleeping had both eyes closed. They may partially or totally open one or both eyes if they are disturbed."

So is our questioner hallucinating? Not necessarily. Chameleons are distinguished not only by their ability to change color but by their knack for moving each eye independently. John E. Simmons, of the American Society of Ichthyologists and Herpetologists, thinks that a shifty chameleon might have been trying to snare our Sarah:

> It is not uncommon for a chameleon to sit motionless for long periods of time with one eye closed and one open, but it is not sleeping when it does this, it is awake and watching for prey items and predators with the open eye.

Simmons also mentions that some lizards have a transparent membrane that closes over the eye to protect it. And according to Richard Landesman, a zoologist at the University of Vermont, some lizards have a pattern of scales and coloration on their eyelids that might fool predators into thinking their eyelids are closed. These lizards might also look like they have one eye closed but are actually capable of seeing with both eyes. Illusions that trick us into thinking they have an eye closed might be used to fool their potential predators as well.

Submitted by Sarah Robertson of Sun Valley, Nevada.

If Grapes Are Both Green and Purple, Why Are Grape Jellies Always Purple?

If you read our first book of Imponderables (entitled, appropriately enough, *Imponderables*), you know how they make white wine out of black grapes. But for those of you who aren't yet fully literate, we'll reiterate: The juice from grapes of any color is a wan, whitish or yellowish hue. White wine is the color of that juice. Red wine combines the juice of the grape along with its skin. The color of the skin, not the juice, gives red wine its characteristic shade.

If you want to prove this premise, go out to your local supermarket or produce store and crush a few red and green grapes into empty containers—you'll be surprised at how similar the colors of the liquids are. But throw grapes with skin into a blender or food processor, and you'll have a very different color. One more suggestion: We recommend using seedless grapes.

Ever since 1918, when Welch's started marketing grapelade, a precursor of the jellies soon to follow, the company used Concord grapes exclusively. By the time the company pioneered the mass distribution of jelly, it was already established in the juice business: Welch's introduced grape juice in 1869, using the same dark Concord grapes. James Weidman III, vice-president of corporate communications for Welch's, told us that the characteristic color of Welch's grape jelly comes from the purple skin of the Concord grape, although the pulp of the Concord grape is white: "Because the basic ingredient in grape jelly is juice, the jelly is therefore purple."

Megan Haugood, account manager at the California Table Grape Commission, told us that consumers now expect and demand the purple color established by Welch's and would be uncomfortable with the appearance of a green grape jelly, even assuming a jelly could be made from green grapes that tasted as good. So all of Welch's competitors adopted the same bright color for their grape jellies.

We asked Sandy Davenport, of the International Jelly and Preserve Association, whether her organization was aware of any green jellies in the marketplace. She started looking through her six-hundred-page directory, finding marketers of rhubarb jelly, partridge berry jelly, California plum jelly, and, of course, kiwifruit jelly. But no green grape jellies in sight. Most of the experts we consulted felt that green grapes are considerably blander in taste and would be unlikely to gain a foothold in the marketplace, even if youngsters could tolerate the idea of eating peanut butter and green jelly sandwiches.

Submitted by Jeff Thomsen of Naperville, Illinois.

Why Didn't Fire Trucks Have Roofs Until Long After Cars and Trucks Had Roofs?

The first fire wagons in America were not motorized. They weren't even horse-drawn. They were drawn by humans. Bruce Hisley, instructor at the National Fire Academy, told *Imponderables* that verifiable records of human-drawn fire engines show that they were in use as late as 1840. Horse-drawn open carriages were then the rule until motorized coaches were introduced in 1912.

By today's standards, the early motorized fire trucks were far from state of the art. Not only were they topless, but they lacked windshields. Early designers didn't realize the environmental impact of greater speeds upon fire crews, especially the driver. Although drivers soon put on goggles to help fight off the wind, rain, and snow to which they were exposed, the addition of windshields was a major advance for safety and driver comfort. So was another afterthought: doors on the sides of the cab.

Of course, it was the rule for the crew to stand on the outside of the truck during runs, exposing them to further unsafe conditions. Although roofs and closed cabs were introduced in the late 1920s or early 1930s, many firefighters continued to ride on the outside. As David Cerull, of the Fire Collectors Club, puts it: "When it comes to change in the fire service, the attitude that prevails is: 'If we did not have them before, why should we have them now?'"

Fire departments did have some legitimate arguments against the covered cabs. Martin F. Henry, assistant vice-president of the National Fire Protection Association, told *Imponderables* that many fire departments, upon receiving covers for their fire trucks, promptly removed them. Why?

> The thinking at the time was that the open cab provided an ability to see the fire building in an unrestricted fashion. Ladder companies, it was thought, would have a better opportunity to know where to place the aerial if they could see the building from the cab.

This objection was particularly strong in urban areas. Cerull points out that when approaching a fire in a highrise, the company could see the upper floors and roofs more easily. Eventually, the "sun roof" solved this problem.

The other complaint about early roofs was that the roofs themselves were fire hazards. Arthur Douglas, of fire equipment manufacturer Lowell Corporation, told us that many of the roofs were protected by weatherproofing, sometimes wood covered with a treated fabric. "Both of these materials were, of course, flammable, thus a high risk on a firefighting unit." The metal roof obviously solved this argument against the closed cab.

So why didn't the changeover occur more quickly? One reason, besides sure inertia, opines Henry, is that the life span of fire trucks is twenty to twenty-five years. Many departments were loath to abandon functioning apparatus. And although many firefighters enjoyed riding the tail-board, statistics accumulated, to no one's surprise, proving that the outside of a truck moving at forty miles per hour wasn't the best place to be during an accident.

But perhaps the precipitating factor in closing cabs is a sadder commentary on our culture than a disregard for safety, as Martin Henry explains:

> Open cabs came to an abrupt halt in major metropolitan areas when it became fashionable to hurl objects at firefighters. All riding positions were quickly enclosed.

Submitted by Scott Douglas Burke of Charlestown, Maryland.

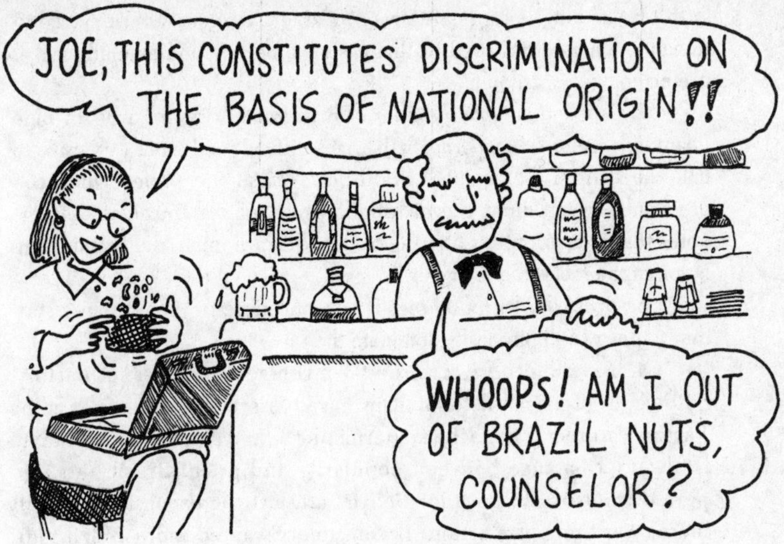

Why Are There So Few Brazil Nuts in Mixed Nuts Assortments?

We contacted a slew of nut authorities and quickly realized that the ratio of various nuts in an assortment is *serious business.* The industry rule of thumb is that there is never less than 2 percent by weight of any type of nut in a mixed-nut assortment, but how do they arrive at the proper proportion? Focus groups, of course. Nut marketers attract roving hordes off the street, sit them down at a table, and find out what people's deepest fears, needs, and fantasies about mixed nut assortments truly are.

And here's why Brazil nuts get the shaft:

1. *People don't like them that much.* If Brazil nuts were popular, you'd see the big nut companies, like Planters, selling whole jars of Brazil nuts. They don't because an insufficient number of consumers would buy them. As one nut expert so accurately put it: "Generally, the last nut remaining in the nut bowl after favorites have been picked by consumers

in the Brazil nut." A representative of Planters Peanuts who preferred to remain anonymous assured us that cashews, pecans, and almonds are all preferred over Brazil nuts.

2. *People don't like the color of Brazil nuts.* Walter Payne, of Blue Diamond, wrote *Imponderables* that it is difficult and quite expensive to take the skin off (i.e., blanch) Brazil nuts. Consumers prefer lighter colored nuts, which limits the distribution ratio of the Brazil nuts. Payne notes that in assortments offering blanched Brazil nuts, "the Brazil population in the mix is much higher."

3. *Brazil nuts are too big.* If you put more of them in an assortment, they would physically dominate the mix.

4. *Brazil nuts are expensive.* We contacted the Peanut Factory, a Rome, Georgia, marketer of many mixed assortments. The folks at the Factory assured us that when determining a ratio of nuts in a mix, there is always a balance between popularity and price. Brazil nuts, imported from Brazil, Peru, or Bolivia, are just too expensive to justify increasing their proportion. If consumers wanted more Brazil nuts, they would complain. But no one we contacted showed the slightest indication that they've ever been confronted with angry consumers demanding more Brazil nuts.

We don't know why you asked this question, Michele, but we hope you are satisfied. If you aren't, I'm sure you derive pleasure from being invited to many parties, where your main role is to eat all the Brazil nuts left by more conventional partygoers.

Submitted by Michele Baerri of Leonie, New Jersey.

Stop the Presses. Days before we submitted this manuscript, we heard from Kimberly J. Cutchins, president of the National Peanut Council:

> As the National Peanut Council, we do not represent other nuts. However, I did call one of our members, Planters, to get a response to this question. A pound of peanuts costs between 35 cents and 70 cents, depending on the size and variety. Brazil nuts cost approximately $1.25 per pound. Although the difference in price of the nuts is significant, it is not the reason there are so few Brazil nuts in mixed nut assortments.

> According to Planters, it is consumer preference that determines the ratio of nuts.

So far, Cutchins merely confirms our research. But then she dropped the bombshell:

> Just recently, Planters has increased the number of Brazil nuts in their assortment to reflect consumer demand.

There you are, Michele. An unprecedented boomlet for Brazil nuts has changed the equation. Can filberts be far behind?

Door-Handle of Fortune
A
B
C
PUSH
CHOOSE A DOOR AND WIN A DREAM KITCHEN, AN EXCITING TRIP TO EUROPE, OR GERMS TO AN EMBARRASSING DISEASE!

Why Do American Doors Have Round Door Knobs, While Many Other Countries Use Handles?

Our correspondents wonder why the knob is virtually ubiquitous when it is much more inconvenient than the handle. Anyone who has ever arrived home carrying three sacks of groceries, with an eight-year-old child and a dog in tow, quickly grasps the notion that it would be far easier to pull down a lever with a loose elbow than to grasp a knob tightly and twist.

All of the hardware sources we contacted indicated that the primary reason knobs are more popular in the United States is simply because customers prefer their looks. Richard Hudnut, product standards coordinator for the Builders Hardware Manufacturers Association, told us that knobs are also easier to manufacture and thus usually cheaper for the consumer.

Joe Lesniak, technical director of the Door and Hardware Institute, says that handles (or "levers," as they are known in the trade) are making a comeback, for at least two reasons. Just as tastes in hemlines vary for no discernible reason, so do preferences for door finishings. As this book is being written, short skirts and door handles are in vogue. But more importantly, the soon to be enforced Americans with Disabilities Act and the Fair Housing Act are going to change Americans' knob habit indefinitely. This legislation mandates that doors in buildings other than one- and two-family dwellings must contain handles and pulls that are easy to activate with one hand and that don't need to be tightly held.

Submitted by Dainis Bisenieks of Philadelphia, Pennsylvania. Thanks also to George Marchelos of 291st BSB.

Why Do Starving Children Have Bloated Stomachs?

How often have we seen pictures of young children, near death from starvation, with emaciated faces and bloated stomachs? The image is haunting and yet ironic: Why do these children, desperately in need of food, have such protruding abdomens?

Bloated abdomens are a symptom of protein calorie malnutrition (PCM). Many of these youths are starving from a generally inadequate calorie consumption and concomitant insufficient protein. But others are suffering from "kwashiorkor," a condition in which children who consume a proper amount of calories are not eating enough protein. Kwashiorkor is most common among many of the rice-based cultures in the third world, where traditional sources of protein (meat, fish, legumes) are uncommon or too costly for the average citizen.

Insufficient protein consumption can lead to severe problems—it produces the lack of energy reflected in the passive, affectless expression of these PCM children. PCM can affect every organ in the body, but it is particularly devastating to the pancreas, liver, blood, and lymphatic system.

A healthy person's blood vessels leak a little fluid, which collects outside of the vessels. Ordinarily, the lymphatic vessels remove this liquid. But when the lymphatic system malfunctions, as it does in PCM children, the fluid builds up in the skin, causing a condition known as edema.

In these children, a particular type of fluid accumulation, ascites (the fluid buildup in the abdominal cavity), accounts for much of the bloated stomach. A little fluid in the abdominal cavity is a desirable condition, because the fluid helps cushion organs. Ascites in isolation may not be dangerous, but they are often a symptom of liver damage. Don Schwartz, a pediatrician at Philadelphia's Children Hospital, told *Imponderables* that membranes often weaken during protein calorie malnutrition, which only adds to the leakage of body fluids into the abdominal cavity.

Dysfunctional livers often swell. The liver is one of the largest organs in the body and usually constitutes 2 to 3 percent of one's entire body weight. According to Dr. Schwartz, an enlarged liver can contribute to a swollen belly.

Of course, not only small children are subject to this condition. Bloated stomachs are common in any individuals suffering malnutrition, and are most often seen in Western countries among sick people who have experienced sudden weight loss. Hospitals are alert to the problem of PCM among adults—one estimate concluded that about 25 percent of hospitalized *adults* in the United States have some form of PCM.

Submitted by Candace Adler of La Junta, Colorado.

Why Do Most Buses and Trucks Keep Their Engines Idling Rather Than Shutting Them Off While Waiting for Passengers or Cargo?

During the first big gas crisis, public service announcements on radio and television urged us not to leave engines idling unnecessarily. Why don't buses and trucks live by the standards we mundane auto drivers do?

The key to the answer is the fuel used in the bus or truck. If you observe carefully, you'll notice that the "idlers" are diesel-powered vehicles. We always thought that bus drivers were leaving engines idle out of laziness, but Morris Adams, of Thomas Built Buses, set us straight:

> These diesel-powered engines require a certain level of heat to operate most efficiently. It is cheaper to leave them running than cold starting. Diesel fuel will last almost twice as long as gasoline when used *under the same atmospheric conditions.*

Idling can also be a safety issue. Most buses, and many big trucks, operate with air brakes. Air brakes can't operate effectively until sufficient air pressure has built up, a process that can take about ten minutes.

And one issue pertains specifically to school or public buses—comfort. Karen E. Finkel, executive director of the National School Transportation Association, explains:

> [Bus riders] want and expect comfort—air conditioning in the summer and heat in the winter. Buses have a massive amount of air space that takes a longer period of time to heat or cool than an individual's automobile.

Submitted by Alka Bramhandkar of Vestal, New York. Thanks also to Brian Dunne of Indianapolis, Indiana.

Why Does One Sometimes Find Sand in the Pockets of New Blue Jeans?

Conceivably, you might have picked up a pair of jeans returned by a previous customer who went to the beach, but that is highly unlikely. We aren't gamblers by nature, but we would be willing to wager a small sum that the jeans in question were stonewashed.

Stonewashed jeans are softened by rubbing against pumice stones during washing. Dori Wofford, a marketing specialist at jeans behemoth Levi Strauss & Co., explains how mighty stones can turn into sand:

> Pumice is a soft white stone that is placed in huge washers along with jeans to be "stonewashed." Pumice stones used in the stonewashing process sometimes disintegrate into tiny particles (or sand) that end up in the pockets of stonewashed jeans.

Obviously, pockets are the one portion of the jeans most susceptible to trapping loose pumice. Any other sand would tend to get rinsed away with the wash water.

Submitted by Lisa R. Bell of Atlanta, Georgia.

Why Do So Many Recreational Vehicle Owners Put Cardboard or Plywood Square Covers over Their Wheels?

The ultraviolet rays from direct sunlight oxidizes the rubber in tires rapidly, leading to premature cracking and drying. Bill Baker, media relations manager for the Recreation Vehicle Industry Association, told us that "there are vinyl covers manufactured and sold to meet this need." We have seen many a custom-cover, emblazoned with family names, mottoes, and mascots, as we passed RVs on the highway. But enterprising and frugal RV owners have found a solution that not only recycles forest products but, not coincidentally, saves them a few bucks.

Submitted by Howard Helman of Manhattan Beach, California.

Why Are Most People Buried Without Shoes?

Personally, we can't imagine an eternal life with uncomfortable shoes. This funeral tradition has always made sense to us; we don't even like to wear shoes when we are awake.

Richard A. Santore, executive director of the Associated Funeral Directors International, reminded us that for those buried in half-open caskets, the shoes (or lack therof) wouldn't be visible and are thus unnecessary. And he added devilishly, aware of the precarious logic behind his assertion, "Now you may ask why we are buried with underwear—I don't know!" We have yet to meet a funeral director without a wicked sense of humor. Comes in handy in the profession, we'd guess.

Several members of the funeral profession claimed that nearly as many folks are buried with shoes as without these days. Indeed, one of the "options" bereaved families are given, when arranging the ceremony, is "burial slippers."

Dr. Dan Flory, president of the Cincinnati College of Mortuary Science, told us that some people believe that a dead person can rest better and "the spirit will not wander if shoes are omitted." But his other theory ties in specifically to the comfort we feel in repressing the finality of death:

> Being without shoes is also a typical sleeping position, and many people like to think of death as a kind of sleep.

We agree with the thesis. But then, following this logic, wouldn't we be buried in pajamas?

Submitted by Harold Fair of Bellwood, Illinois. Thanks also to N. Dale Talkington of Yukon, Ohio, and Deone Pearcy of Tehachapi, California.

Typical NEW YORK Resident:
CHEWS 2 PACKS OF GUM DAILY
HAS NEVER BEEN TO STATUE OF LIBERTY
CALLS THE AVENUE OF THE AMERICAS "SIXTH AVENUE"
CAN RIDE SUBWAY STANDING UP WITHOUT HOLDING ON
HAS NEVER EATEN A NEW YORK STRIP STEAK
~NEVAH EV'N HOIDA IT!

What Is a "New York Steak"? Is It a Cut of Meat? Is It a Part of the Cow? And Why Can't You Find "New York Steaks" in New York?

Reader Douglas Watkins, Jr., posed this Imponderable four years ago and has waited patiently while we've been researching it ever since. We've contacted about twenty meat experts and have come to the conclusion that Douglas might decide, after having read this chapter, that the wait wasn't worth it.

One of our first conversations on the topic was with Merle Ellis, a combination butcher-media star, who wrote an excellent book, *Cutting-up in the Kitchen,* which, among other things, rigorously defines all of the cuts of a cow. Ellis places the New York steak in the short loin, between the rib and the sirloin of the cow, in the middle of the back. The short loin contains a high percentage of fat and very little muscle. As a result, the most expensive cuts of meat come from this 10 percent of edible beef: filet mignon, porterhouse, tournedos, T-bones, and, alas, the New York steak. But what part of the short loin constitutes the New York steak?

On this question, we're afraid, the National Association of Hotel and Restaurant Meat Purveyors, the American Meat Institute, the National Live Stock & Meat Board, and the numerous restaurateurs and butchers we spoke to cannot agree. Merle Ellis votes for the top loin muscle of the tenderloin; others vote for the center of the loin. Indeed, the *New York Times*'s Molly O'Neill was inspired by this dilemma and wrote a whole story about it in the January 2, 1991, article "In Search of New York Steak? Ask Anywhere but New York." She eventually came to the same conclusion that we have: "There is no part of a cow with New York stamped on it, nor any particular cut of beef that is peculiarly 'New York.'"

Several of our sources, including Dr. Stuart Ensor, of the National Live Stock & Meat Board, suggested that "New York" does not refer to a particular type of cut but is merely a promotional adjective. New York equals "the best." This helps explain the last part of our Imponderable: why New York steaks are rarely called "New York steaks" in New York! Alas, the grass is always greener on the other side. While New York steaks might have allure in the West, New Yorkers conjure the Midwest as the true home of the stockyards and therefore the best beef. As Ellis so eloquently put it in *Cutting-up in the Kitchen:*

> The top loin muscle becomes a New York Strip in Kansas City and Kansas City strip in New York City. . . . On your side of the street it could be almost anything: Shell Steak, Hotel Steak, Sirloin Club Steak, Boneless Club Steak or Charlie's Gourmet Special. Whatever it's called, it too will be most definitely on the expensive side.

Submitted by Douglas Watkins, Jr., of Hayward, California.

Why Are Some Aluminum Cans—Even Different Cans of the Same Product—Harder to Crush Than Others?

Let's get one issue out of the way immediately. Even the largest beverage producers (or "fillers," as they are known in the canning industry) don't make their own cans. A company like Coca-Cola has many different suppliers. And these canners may have many plants designed to produce cans of slightly different composition and dimensions.

Jeff Solomon-Hess, executive editor of *Recycling Today*, reminds us that not all beverage cans are made of aluminum only:

> Manufacturers commonly produce two types of beverage cans today: all aluminum and bi-metal. The all aluminum can crushes slightly easier because of its thinner construction. The bi-metal can consist of a steel body and an aluminum top. While steel technology allows for much thinner construction than in the past, the steel cans remain slightly thicker and harder to crush.

William B. Frank, of the Alcoa Technical Center, points out that the "trained consumer" can differentiate between the two types of cans:

> The steel can has a matte finish on the bottom and a "stiff" sidewall; the aluminum can is shiny on the bottom and is less stiff after opening. A sure way to distinguish a steel can from an aluminum can is to see whether a small magnet will stick to the body or bottom of the can. Ferromagnetism is used to separate steel cans from aluminum cans in recycling of UBCs [used beverage cans].

We don't mean to imply that all aluminum cans are created equally. Indeed, off the top of his head, E.J. Westerman, manager of can technology at Kaiser Aluminum, listed five reasons why all-aluminum cans might vary in crushability:

> 1) There are small differences in diameters of cans (e.g., Coors-type cans are 2 9/16", while most other American cans are 2 11/16").

2) Fillers give aluminum producers different requirements for alloy strength and composition.

3) Aluminum producers vary the wall thicknesses of the cans from about 0.0039" to approximately 0.0042".

4) Cans can vary in "wall thickness gradation" at the neck and base of the can, which can affect the point at which "column buckling" begins when a can is crushed.

5) The end or lid diameter and thickness can have a great impact upon crushability, particularly if the can is not crushed in the axial direction.

Solomon-Hess feels that the role of the crusher cannot be underestimated:

> Most people place the can on end and step on it. If you step on it with a rubber sole shoe (such as a sneaker), your foot sometimes forms an airtight seal. This results in the inside air pressure of the can making it more rigid, and again, harder to crush. Tilting your foot slightly and allowing the air to escape or laying the can on its side makes it crush much easier.

And don't forget the condition of the can itself. We have found ourselves, often unconsciously, squeezing or popping aluminum cans. As soon as you mess with the structural integrity of the container, it becomes much easier to crush. Solomon-Hess even offers an experiment that you can do at home (children under eighteen, please ask your parents for permission); for reasons too obvious to explain, this experiment may best be conducted outdoors:

> If you put a couple of small dents in the can, it requires much less pressure to crush, so people who routinely "pop" the sides of cans as they drink from them find the empty cans easier to crush. For a dramatic example, try this experiment.
>
> Take an empty soda can and place it upright on a hard floor (concrete works best) and ask someone who weighs under 150 pounds to very carefully place his or her full weight on the can with one foot and balance there. Most cans will support such a weight. Now take two pencils and simultaneously tap opposite sides of the can with a firm motion. The

resulting dimples cause the can to quickly collapse by creating weak spots in the can's structure. The effect is the same as scoring a sheet of glass in order to break it cleanly.

Not quite the same—you don't need a sponge to clean up after breaking a sheet of glass.

Submitted by Roy L. Youngblood of Oceanside, California.

Why Are Pigs Roasted with an Apple in Their Mouth?

Is this the first reference book in history to devote three chapters to the appearance of cooked pork products? Maybe so, but we follow the lead of our readers, and pig pulchritude seems to be very much on your minds.

Unfortunately, we have not gotten to first base in answering how this practice originated. But the same pork experts who guided us before are quick to assure us that if an apple was put in the suckling pig's mouth before roasting, it would quickly turn to a texture more like apple sauce than apples. The pristine apple is put in the pig's mouth after it is removed from the spit.

So the apple is purely a decorative item, perhaps inserted out of a sense of loss of the aforementioned checkerboard pattern on hams. Tonya Parravano, of the National Live Stock & Meat Board, told *Imponderables* that most recipes

> suggest placing a small block of wood in the pig's mouth before roasting to brace it so an apple can be inserted later. Cranberries or cherries are often used in the eye sockets and some like to give the pig a collar of cranberries, parsley, or flowers.

The poor pig, whose very name conjures up filth and sweat in common parlance, is subjected to even greater indignities before it is consumed.

Submitted by Christine Whitsett of Marion, Alabama.

Why Do Many Streets and Sidewalks Glitter? Is There a Secret Glittery Ingredient?

Two main ingredients create the glittering appearance of our concrete and asphalt roadway surfaces: natural rocks and glass. When it comes to asphalt and concrete, the contents are always hybrids, mixes of stones, sand, petroleum derivatives, and "fillers," ingredients that aren't necessary for the integrity of the pavement but provide bulk—the construction equivalent of Hamburger Helper.

Sand, glass, silicon, and many natural stones, such as quartz, all glitter. According to Jim Wright, of the New York State Department of Transportation, glass is included in roadways as a way of recycling used byproducts. Other fillers, such as used tires, are thrown in to the mix as a way of keeping solid waste dumps from looking like canyons exclusively devoted to showcasing beaten-down, giant chocolate doughnut lookalikes.

Construction engineers are sensitive to the aesthetics of streets and sidewalks. Billy Higgins, director of congressional relations at the American Association of State Highway and Transportation Officials, told *Imponderables* that glass is often used primarily to make surfaces more attractive. For example, Higgins says that concrete sidewalks routinely are smoothed down with a rotary blade to allow the shiny surfaces to show off to best effect.

Portland cement, a mix made primarily of limestone and clay, becomes a particularly glittery surface when it bonds with sand and other filler agents to become concrete. Portland cement concrete becomes shinier and shinier as it is used, as Thomas Deen, executive director of the Transportation Research Board of the National Research Council, explains:

> Some of the aggregate used in portland cement concrete is like a natural mirror; that is, it reflects light. In theory, all aggregate in concrete is completely coated with cement. However, the aggregate on the very top surface of the street or sidewalk will lose part of that coating due to the weathering and vehicular or pedestrian traffic. Once exposed, the light from the sun, headlights, street lights, or other sources bounces off the tiny surfaces of the aggregate, causing the streets and sidewalks to glitter.

Submitted by Sherry Spitzer of San Francisco, California.

Why Hasn't Beer Been Marketed in Plastic Bottles Like Soft Drinks?

Now that the marketing of soda pop in glass bottles has pretty much gone the way of the dodo bird, we contacted several beer experts to find out why the beer industry hasn't followed suit. The reasons are many; let us count the ways:

1. Most beer sold in North America is pasteurized. According to Ron Siebel, president of beer technology giant J.E. Siebel Sons, plastic bottles cannot withstand the heat during the pasteurization process. Plastics have gained in strength, but the type of plastic bottle necessary to endure pasteurization would be quite expensive.

John T. McCabe, technical director of the Master Brewers Association of the Americas, told *Imponderables* that in the United Kingdom, where most beer is not pasteurized, a few breweries are marketing beer in plastic bottles. Siebel indicated that he would not be surprised if an American brewer of nonpasteurized (bottled) "draft" beer doesn't try plastic packaging eventually.

2. Breweries want as long a shelf life as possible for their beers. According to Siebel and McCabe, carbon dioxide can diffuse through plastic and escape into the air, while oxygen can penetrate the bottle, resulting in a flat beverage. Glass is much less porous than plastic.

3. Sunlight can harm beer. Siebel indicates that beer exposed to the sun can develop a "skunky" taste and smell; this is why many beers are sold in dark and semi-opaque bottles.

4. Appearance. We have no market research to support this theory, but we wouldn't guess that beer would be the most delectable looking beverage to the consumer roving the supermarket or liquor store aisle.

One might think that breweries would kill to package their product like soft drinks. Could you imagine the happy faces of beer executives as they watched consumers lugging home three-liter plastic bottles of suds?

Submitted by John Lind of Ayer, Massachusetts.

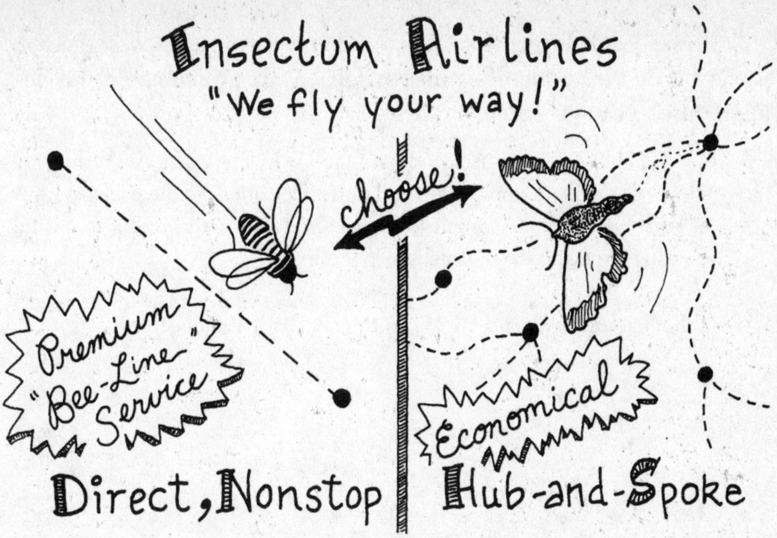

Why Do Some Insects Fly in a Straight Line While Others Tend to Zigzag?

As entomologist Randy Morgan of the Cincinnati Insectarium puts it, "Flight behavior is an optimization of the need to avoid predators while searching for food and mates." Gee, if Morgan just eliminated the word "flight" and changed the word "predators" to "creditors," he'd be describing *our* lives.

Notwithstanding the cheap joke, Morgan describes the problems of evaluating the flight patterns of insects. An insect might zigzag because it is trying to avoid an enemy or because it doesn't have an accurate sighting of a potential food source. A predatory insect might be flying in a straight line because it is unafraid of other predators or because it is trying to "make time" when migrating; the same insect in search of food might zigzag if its target wasn't yet selected.

Leslie Saul, Insect Zoo director at the San Francisco Zoological Society, wrote *Imponderables* that the observable flying patterns of different insects can vary dramatically:

> Flight paths are usually determined by visual, auditory, or olfactory stimulation. For example, bees and butterflies orient to the color and size of flowers; dragonflies orient to their prey items; moths orient to a wind carrying a specific smell, usually a "pheromone."

Submitted by Dallas Brozik of Huntington, West Virginia.

What's the Difference Between "French" and "Italian" Bread?

Not a whole lot, it turns out. But there are enough differences in ingredients to account for the subtle differences in taste and, particularly, texture.

Baking consultant Simon Jackel kindly wrote us a primer on the subject:

> French and Italian breads are made from the same basic ingredients: flour; water; salt; and yeast. Both use "strong" flours. And they both develop crisp crusts in the oven due to the injection of live steam.
>
> But there the similarity ends, because "French" breads, but not "Italian," also incorporate small amounts of shortening and sugar in the

formulation. The effect of these additional ingredients is to allow the French dough to expand more and become larger in volume, lighter in consistency, and more finely textured in the interior. In contrast, Italian breads are denser and less finely structured in the interior.

The shape of the loaf may tip off the nationality of the bread. Sometimes, "Italian" bread is formed in a football-like shape, as opposed to the sleeker "French." And sometimes "Italian" bread is topped with sesame seeds, an embellishment that would probably make the French pop their berets.

Submitted by Todd Kirchmar of Brooklyn, New York.

Why Have So Many Pigeons in Big Cities Lost Their Toes?

The three main dangers to pigeons' toes are illnesses, predators, and accidents. Pigeons are susceptible to two diseases that can lead to loss of toes: avian pox, a virus that first shrivels their toes to the point where they fall off, and eventually leads to death; and fungal infections, the price that pigeons pay for roaming around in such dirty environments.

Nonflying predators often attack roosting pigeons, and the toes and lower leg are the most vulnerable part of pigeons' anatomy. Steve Busits, of the American Homing Pigeon Fanciers, told *Imponderables* that "Rats or whatever mammal lives in their habitat will grab the first appendage available."

Accidents will happen, too. Busits says that toes are lost in tight spaces, namely "any cracks or crevices that their toes can become stuck in." Bob Phillips, of the American Racing Pigeon Union, adds that toes get lost while pigeons are in flight, with television antennas and utility wires being the main culprits.

Submitted by Nancy Metrick of New York, New York. Thanks also to Jeanna Gallo of Hagerstown, Maryland.

How Do Highway Officials Decide Where to Put a "Slippery When Wet" Sign?

The holy grail of signage policy is the *Manual on Uniform Traffic Control Devices,* a Federal Highway Administration publication that is followed by state jurisdictions as well. In other *Imponderables* books, we've regaled you with complex descriptions of how the MUTCD specifies exactly where, how, and why certain traffic signs should be posted.

But the MUTCD passage on "Slippery When Wet" signs is remarkably vague by comparison:

> The Slippery When Wet sign is intended for use to warn of a condition where the highway surface is extraordinarily slippery when wet.
>
> It should be located in advance of the beginning of the slippery section and at appropriate intervals on long sections of such pavement.

Without specific instructions, state highway agencies have to decide where to place signs.

So how do they decide what roadways are slippery?

1. According to Harry Skinner, chief of the traffic engineering division of the Federal Highway Administration's Office of Traffic Operations,

> Highway surface will become extraordinarily slippery if the aggregate or rock in the pavement becomes polished and cannot drain off all water with which it comes in contact.

Obviously, all surfaces are more slippery when wet than when dry, and a roadway shouldn't be slapped with a "Slippery" sign merely because it becomes slick when ice accumulates. In fact, specific "Icy Pavement" signs are available to warn about these conditions. Sometimes, "Slippery" signs are erected precisely because the roadway looks innocuous; one state document we read indicated that a "Slippery When Wet" sign should be placed where "skid resistance is significantly below that normally associated with the particular type of pavement, or where there is evidence of unusual wet pavement."

2. Bridges tend to be more slippery than adjacent pavements and may warrant a sign.

3. If a roadway is suspected of being slippery, engineers can do a technical analysis, determining the "coefficient of friction.

4. But the most common motivation for placing a "Slippery When Wet" sign is a little more depressing, as Joan C. Peyrebrune, technical projects manager of the Institute of Transportation Engineers, explains:

> Generally, signs are placed at locations where an accident analysis indicates that a significant number of accidents caused by slippery conditions has occurred. The number of accidents that warrant a "Slippery When Wet" sign varies for each state.

This strategy reminds us of an old cartoon we found in a sick joke book: As an automobile pileup of epic proportions turns an intersection into a scrap-strewn catastrophe, an expressionless policeman mounts a ladder to place an "out of order" sign over a failing traffic light.

The intent of the "Slippery When Wet" sign is no different from the "Falling Rock" sign we talked about in *When Did Wild Poodles Roam the Earth?* The hope is that the driver, fearing impending doom, will slow down a tad. And maybe now that you know that these notices serve as markers for misguided drivers who once veered off the road, the signs will do their jobs even more effectively.

Submitted by Herbert Kraut of Forest Hills, New York.

Can One Spider Get Caught in the Web of Another Spider? Would It Be Able to Navigate with the Skill of the Spinner?

Yes, spiders get caught in the webs of other spiders frequently. And it isn't usually a pleasant experience for them. Theoretically, they might well be able to navigate another spider's web skillfully, but they are rarely given the choice. Spiders attack other spiders, and, if anything, spiders from the same species are more likely to attack each other than spiders of other species.

Most commonly, a spider will grasp and bite its intended victim and inject venom. Karen Yoder, of the Entomological Society of America, explains, "Paralysis from the bite causes them to be unable to defend themselves and eventually they succumb to or become a meal!"

Different species tend to use specialized strategies to capture their prey. Yoder cites the example of the Mimetidae, or pirate spiders:

> They prey exclusively on other spiders. The invading pirate spider attacks other spiders by luring the owner of the web by tugging at some of the threads. The spider then bites one of the victim's extremities, sucks the spider at the bite, and ingests it whole.
>
> The cryptic jumping spider will capture other salticids or jumping spiders and tackle large orb weavers in their webs. This is called web robbery.

Other spiders will capture prey by grasping, biting, and then wrapping the victim with silk. Leslie Saul, Insect Zoo director of the San Francisco Zoo, cites other examples:

> Others use webbing to alert them of the presence of prey. Others still have sticky strands such as the spiders in the family Araneidae. Araneidae spiders have catching threads with glue droplets. The catching threads of Uloborid spiders are made of a very fine mesh ("hackel band"). *Dinopis* throws a rectangular catching web over its prey item and the prey becomes entangled in the hackle threads.

Saul summarizes by quoting Rainer F. Foelix, author of *Biology of Spiders*: "The main enemies of spiders are spiders themselves."

Not all spiders attack their own. According to Saul, there are about twenty species of social spiders that live together peacefully in colonies.

Submitted by Dallas Brozik of Huntington, West Virginia.

Over the years, we have received many Imponderables about McDonald's but have found answers elusive. Now, with the help of Patricia Milroy, customer satisfaction department representative, we can finally unburden you of some of your obsessions.

To Exactly What Is McDonald's Referring When Its Signs Say "Over 95 Billion Served"?

This Imponderable has provoked more than one argument among our peers. We're proud to finally settle the controversy. No, "95 billion" does not refer to customers served, sandwiches served, food items served, or even hamburgers served.

The number pertains to the number of *beef patties* served. A hamburger counts as one patty. A Big Mac counts as two. A quarter-pounder with cheese counts as one. A double cheeseburger counts as two. Got it? The practice undoubtedly started when McDonald's served no other sandwiches besides (single-patty) hamburgers and cheeseburgers.

By the time you read this, McDonald's will have "turned over" the sign and added another digit. Expectations are that the corporation will have sold 100 billion beef patties before the end of 1993.

Submitted by Jena Mori of Los Angeles, California.

Why Are McDonald's Straws Wider in Circumference Than Other Restaurant or Store-Bought Straws?

McDonald's has test-marketed numerous sizes and materials for their straws. Milroy says:

> After working with our suppliers and testing them with our customers, we've found that the present size of our straws is preferred by the majority of our customers.

One of the readers who posed this question guessed the key to the wider circumference of McDonald's straws—many, many milkshakes get sold at the Golden Arches. Any fan knows of the frustration of trying to suck up a thick glop of milkshake through a narrow straw: liquid gridlock. Sure, Mickey D's might give us more straw than we need for Coca-Cola, iced tea, or milk, but when we choose milkshakes, the bigger the better.

Submitted by Melanie Dawn Parr of Baltimore, Maryland. Thanks also to Sandra Baker of Nicholasville, Kentucky.

Why Are the Burgers Upside-Down When You Unfold the Wrapper of a McDonald's Hamburger?

For the same reason you put a gift upside down before you wrap it. As Milroy puts it:

> To provide a neat appearance, a hamburger is placed upside-down on its wrapper, then the ends of the wrapper are brought together on the bottom side of the hamburger. The hamburger is placed right side up in the transfer bin for sale.

If the burger was placed on the wrapper upright, the "loose ends" of the wrapping paper would land atop the finished product, giving it an unkempt appearance and threatening the unraveling of the paper. Using the preferred method, the loose ends of the wrapping paper end up on the bottom of the wrapped burger as it is put in the bin for sale, allowing gravity and the weight of the burger to hold the loose ends in place.

Submitted by Renate Dickey of Macon, Georgia.

What, Exactly, Is the McDonald's Character "The Grimace" Supposed to Be?

Milroy reported that Imponderables readers are not alone; this is among the most asked questions of the corporation. What does this say about our culture?

We're not here to judge, however, so we are proud to announce the official position of McDonald's on the exact description of The Grimace: "He is a big fuzzy purple fellow and Ronald's special pal." That's it. Regardless of our prodding, our cajoling, our penetrating interrogation, our rare paroxysms of hostility, this was the most we could prod out of our golden-arched pals. But we are assured that this is as much as Ronald McDonald himself knows about his fuzzy purple friend.

Submitted by Michael Weinbeyer of Upper Saint Clair, Pennsylvania. Thanks also to Joe Pickell of Pittsburgh, Pennsylvania; Samuel Paul Ontallomo of Upper Saint Clair, Pennsylvania; Nicole Cretelle of San Diego, California; Ruth Homrighaus of Gambier, Ohio; and Liam Palmer and Jonathan Franz of Corbett, Oregon.

What Did Barney Rubble Do for a Living?

We have received this Imponderable often but never tried to answer it because we thought of it as a trivia question rather than an Imponderable. But as we tried to research the mystery of Barney's profession, we found that even self-professed "Flintstones" fanatics couldn't agree on the answer.

And we are not the only ones besieged. By accident, we called Hanna-Barbera before the animation house's opening hours. Before we could ask the question, the security guard said, "I know why you're calling. You want to know what Barney Rubble did for a living. He worked at the quarry. But why don't you call back after opening hours?" The security guard remarked that he gets many calls from inebriated "Flintstones" fans in the middle of the night, pleading for Barney's vocation before they nod off for the evening.

We did call back, and spoke to Carol Keis, of Hanna-Barbera public relations, who told us that this Imponderable is indeed the company's most frequently asked question of all Flintstone trivia. She confirmed that the most commonly accepted answer is that Barney worked at Fred's employer, Bedrock Quarry & Gravel:

> However, out of 166 half-hours from 1960–1966, there were episodic changes from time to time. Barney has also been seen as a repossessor, he's done top secret work, and he's been a geological engineer.
>
> As for the manner in which Barney's occupation was revealed, it was never concretely established (no pun intended) [sure]. It revealed itself according to the occupation set up for each episode.

Most startling of all, Barney actually played Fred's boss at the quarry in one episode. Sure, the lack of continuity is distressing. But then we suspend our disbelief enough to swallow that Wile E. Coyote can recover right after the Road Runner drops a safe on Coyote's head from atop a mountain peak, too.

Hanna-Barbera does not have official archives, so Keis couldn't assure us that she hadn't neglected one of Barney Rubble's jobs. Can anyone remember any more?

Submitted by Rob Burnett of New York, New York.

For any other readers who submitted this Imponderable, please write so that your name can be included for future editions.

Why Do We Wave Polaroid Prints in the Air After They Come Out of the Camera?

Imponderables often has to ask anthropologists, anatomists, physiologists, or geneticists questions about bodily quirks or anachronisms. Why do we have patches of hair between our knuckles? Why do we have an appendix? What good are earlobes? More often than not, experts shrug their shoulders and reply that at one time in our evolution these features might have served some purpose, but their function is now lost in obscurity. Humans have many such vestigial anatomical remnants.

Likewise, some human activities that now seem meaningless might have served some purpose in an earlier era. To wit: gratuitous hand flapping. We have all seen someone inadvertently consume something that was too hot to put in the mouth. What is the universal cure for a scorched throat? Invariably we see the victim waving a hand violently up and down in front of an opened mouth.

How can flapping your limbs possibly solve the problem of a 900-degree pizza hitting the roof of your mouth? It can't, of course. We can only surmise that in the paleolithic period, perhaps a now-extinct flying insect was preternaturally attracted to burning mouth flesh, and that this waving of hands served as a deterrent.

Other examples of unproductive flapping are not confined to empty hands, however. As our astute field observer, Christine Schomer, points out, millions of amateur photographers can be seen flapping just-issued Polaroid prints with the enthusiasm of Chubby Checker demonstrating The Fly, a dance whose moves might have been inspired by instant photographers waiting for their pictures to develop.

We contacted the folks at Polaroid to ask if there is any method to the seeming madness of flappers. Bob Alter, in the Public Affairs and Community Relations department at Polaroid, told us that flapping undeveloped prints doesn't serve any useful function whatsoever. In fact, if the prints are waved too vigorously, the picture will bend.

But unlike earlobes or finger hair, at least the Polaroid flapping can be explained. Many years ago, Polaroid prints used to come out of the camera in a two-part sandwich, with the positive print sticking to the negative. The two parts were peeled apart. Often, a little polymer, the agent used to transfer the negative to the positive print, stuck to the positive print.

Photographers often waved the print in order to dry off the tacky polymer, in the mistaken belief that the photograph would develop sooner. Now Polaroid uses "integral film," so that the print comes out of the rollers in a self-contained unit. The transfer from negative to positive occurs inside the unit. so that the exposed print isn't moist. Flapping the print around does nothing except kill time until the photo is developed.

So Polaroid flapping is a perfect example of vestigial behavior, an activity that once had some justification and now has none. When will this characteristic be bred out of us? Is Polaroid flapping motivated by genetic or environmental causes? Nature or nurture? Stay tuned.

Submitted by Christine Schomer of New York, New York.

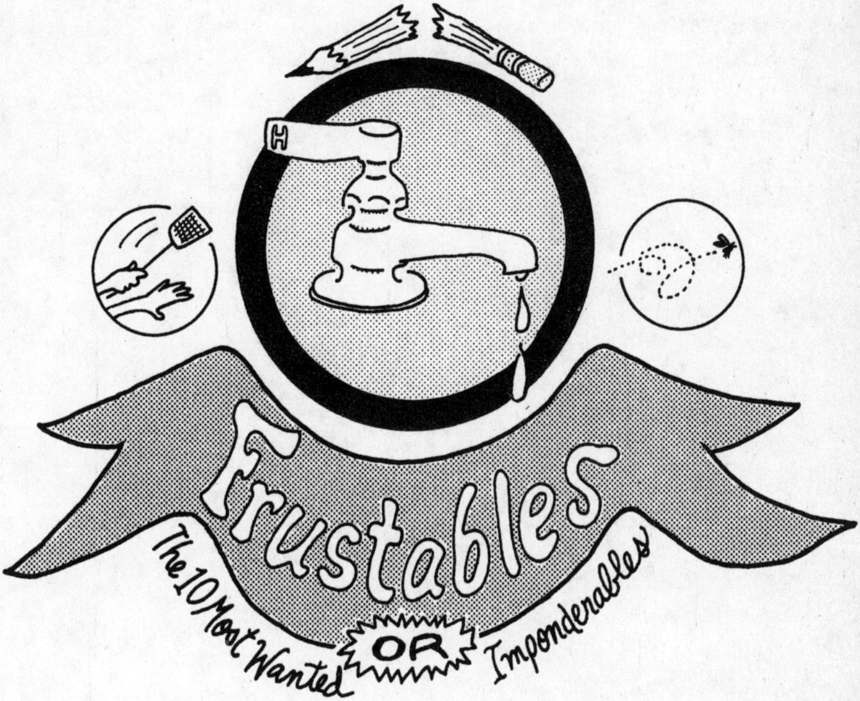
H
Frustables
The 10 Most Wanted
OR
Imponderables

Impressed with our perspicacity in the Imponderables section? Whenever we get full of ourselves, gloating about how we stumbled upon the solution to a knotty Imponderable, our darker side whispers, "Frustables. Frustables. Frustables."

Ah, Frustables (short for "Frustrating Imponderables"), the ten Imponderables we most wanted to answer for this book but could not. In most cases, we contacted experts who came to no useful consensus, or we suspect our sources are not delving into the Imponderable with the depth it deserves.

So we leave it to you. Can you help? We offer a complimentary autographed copy of our next volume of *Imponderables* to the reader who supplies the best answer to each, or the first reader who leads us to the proof that supplies the answer. And of course, your contribution will be printed and acknowledged in the book. Remember, only you can prevent Frustability.

FRUSTABLE 1: *Why do we close our eyes when we kiss?*

Interested in any and all theories, as well as any amusing anecdotes about this subject.

FRUSTABLE 2: *Why do women "of a certain age" usually start wearing their hair shorter?*

The hairdressers and beauty consultants we spoke to, for the most part, felt there was no good reason for women to wear their hair shorter as they got older. Why has ever-shortened hair become a traditional fashion statement of mature women? Or is this custom merely a way to avoid the inconvenience of dealing with long hair?

FRUSTABLE 3: ***Why do the clasps of necklaces and bracelets tend to migrate from the back toward the front?***

Can this phenomenon be explained by some weird, unwritten rule of physics? Does the slight extra weight of the clasp affect its ability to stay put on the wrist or neck? We have received this complaint from several women and so far have heard no good explanation.

FRUSTABLE 4: ***Why is it customary to include the full address of the recipient of a business letter before the salutation?***

Presumably, the recipient of a letter knows his or her address. When a letter is prepared for a window envelope, this practice saves the sender the time and aggravation of addressing an envelope. But isn't it a waste of time and space otherwise?

FRUSTABLE 5: ***Why do most women like shopping more than men?***

Yes, we *know* this is a gross generalization. But it is fair to say that far more women were "born to shop" than men. We thought men were supposed to be the hunter-gatherers. Why do women seem to derive far more psychic benefits from shopping than men?

FRUSTABLE 6: ***How and why did the association between wearing eyeglasses and nerdiness and/or greater intelligence begin?***

Was the original assumption that people needed glasses because they wore out their eyes reading books? Were glasses considered to be signs of physical weakness? This is one stereotype that never made much sense to us.

FRUSTABLE 7: ***Why and where did the tradition of tearing down football goalposts begin?***

We're confident that the practice started during the days of early American college football, but our usually reliable college football experts can't pinpoint either where it started or whether it began as a demonstration of joy at a victory or a riot after a defeat.

FRUSTABLE 8: ***Why do artists, models, and bohemians wear black clothing?***

In almost any metropolitan area in the Western world, black seems to be a uniform of hipness. Sure, we know that "black makes you look thinner," "everybody looks good in black," "black goes with everything," etc. But why would folks who wouldn't think twice about dyeing their hair purple or inserting earrings through any possible bodily protuberance find black eternally chic?

FRUSTABLE 9: ***Why is the best restaurant coffee better than home-brewed coffee?***

We expect to hear complaints about the premise of this question. Yes, there is plenty of pitiful coffee served in restaurants and cafés, but several readers have asked why they can't make coffee at home that competes with the best restaurant coffee, especially when they are often using more expensive coffee beans than the restaurant.

FRUSTABLE 10: ***Why don't women spit more?***

Yes, we know we are culturally conditioned to consider spitting to be the domain of uncouth males. We realize that males of yore smoked cigars and chewed tobacco that stimulated saliva production. But most of the tobacco-abstaining, ultracouth males we've spoken to commonly feel the urge to spit.

Yet many women deny they ever have such a need. Can there be a physiological explanation? Are women suppressing a desire to spit?

Can you help us? We have great expectorations about this Frustable!

Frustables Update

Our Readers Respond to the Frustables First Posed in *When Did Wild Poodles Roam the Earth?*

FRUSTABLE 1: ***Why do women often go to the restroom together? And what are they doing in there for so long?***

Since we are in the rare position of having read several hundred people, mostly female, describe their restroom habits in graphic detail, and since we are about to embark on a dissection of such habits in detail worthy of a Ph.D. in anthropology, it is only fair we share with women what a men's room is like: a morgue.

About the only place where one can observe the average man more uptight is in an elevator. Even if a male should happen to saunter into a bathroom with a friend, any vocal interplay between them is strictly prohibited (unless approved in writing by Major League Baseball, of course). Men glide over to the urinals, unzip, and look straight ahead with a steadiness of gaze that a marine drill sergeant would envy. They go about their business with the seriousness and speed of someone paying by the hour. By our casual observation, an embarrassing percentage wash their hands as they make a beeline for the exit door. Average time in restroom? Fifteen seconds. Maybe we're exaggerating a tad, but you get the idea.

No wonder, then, that men are amazed that women will troop off together to the bathroom, just as the dinner table is engrossed in a fascinating conversation about the relative merits of the Tampa Bay Buccaneers and the Phoenix Cardinals, or about whether the family auto requires 10–40 or 10–50 motor oil. As their dinner curdles, the men wonder what happened to the women. Have they been spirited into UFOs? Having a Tupperware party? Having a sauna? Here's what we heard:

Safety

We were surprised how many women echoed the sentiments of Lin Sherfy of Nevada, Iowa:

> Practically from birth, Mommy tells her little girl how dangerous public restrooms are. They are dirty, they are filled with strange, unspecified diseases, and they teem with evil strangers who must never be spoken to. One of the most popular urban legends, after all, deals with the little girl who was drugged by an evil woman in a department-store restroom, her hair dyed. Her mother "just happened to catch a glimpse of her shoes as she was carried out the door and rescued her!"
>
> Consciously or subconsciously, a woman feels vulnerable with half her clothes down around her ankles and only half a door between her and whomever walks into the restroom. If there's another woman around, most women will issue a general invitation and a group will go together to watch out for each other. Besides, it's nice to have someone to hold the stall door shut since the lock has almost always been removed, and to pass you a Kleenex when, as usual, there isn't any toilet paper.

Meryl Silverstein of Brooklyn, New York, also mentioned that women are conditioned to fear restrooms from an early age:

> Despite many people insisting that things are worse "now" than they were "then," when I was in elementary school, girls were never permitted to leave the classroom singly, certainly not to go to the bathroom. Presumably a would-be mugger or child molester would be deterred by two little girls. The college I attended had signs on the restroom doors advising women, "Never enter a bathroom alone." Old habits die hard.

Several women mentioned that they fear not only physical attacks but unwanted attention from lecherous men en route to the bathroom. We received a particularly biting primer on the subject from Nancy Tropkoff of Brunswick, Ohio, who even included suggested dialogue to assist, as she so delicately puts it, "moron-evaders":

> Why would you want to go alone if you could be molested by some moron? [If you run into a jerk], your companion can say, "Let's go, I gotta go real bad!" In especially urgent times, "Mrphh-gag-gag . . . I don't feel so good" is also effective. On the way back from the bathroom, friends can also be a human shield ("No, I haven't seen her.") If a woman is somewhere with husband and kids, she will enlist children ("I know you have to go. At least *try* to go!").

To Kill Time

High on the list of most women's pet peeves are the long lines to get into restrooms. Diane Larson of Lakeville, Minnesota, summed up the thoughts of many women when she said that having a companion along is preferable so "that you have someone to talk to during the interminable wait to get into a stall."

Hair/Make-Up

This may surprise our male readers, but beauty makeovers do not occur only in salons and on afternoon TV talk shows—actually, the epicenter of fashion and cosmetics is not Paris or New York City but your local women's room. For men, a toilet is a toilet. For women, evidently a toilet is a toilette, or more precisely, a continuing education course in toilette, as Lin Sherfy explains:

> Any group of more than two women invariably includes one who's recently read one of those "How To Make Lipstick Stay On" magazine articles, and often she's also read that "everyone ought to use a lipstick brush." So there stand her companions, waiting patiently because it would be rude to walk out and leave a friend alone in that evil place, while she gets out her lipstick brush, unscrews brush and lipstick, loads the brush, carefully paints her lips, blots, powders the blotted lipstick, picks her brush up off the floor where it rolled, washes it, dries it, loads the brush, paints over the powdered lipstick, blots . . . and then, she takes her comb and brush and spritzes spray. . . "

You get the idea.

Sharon Brandon of Indianapolis, Indiana, wrote:

> Women are (wrongly) conditioned by our culture to place a heavy emphasis on their appearance. Therefore they feel a need to make sure they still look "all right" at some point in the evening.

All the respondents who mentioned this point added that men are "easy graders" in this department. As Kelli Zimmerman of Milwaukee, Wisconsin, put it,

> My date isn't going to tell me that something doesn't look right. My girlfriend will.

Some women use their companions for positive feedback, too, such as Joan Cartan-Hansen of Boise, Idaho:

> Women seem to need confirmation that they look okay. After fixing your face in the bathroom mirror, it is nice to turn to your companion for a quick approval.

Female Bonding

We heard from many fans of Deborah Tannen's bestseller, *You Just Don't Understand.* Tannen posits that while men are task-oriented in word as well as deed, what she calls "rapport talk" is extremely important to women. Mary Roush of Wilmington, Delaware, explains:

> Where better to have conversational intimacy than in a women-only restroom? . . . You have to go to the restroom together to help form a relationship or otherwise miss an alliance-building, affiliative opportunity. The task (going to the restroom) is simply the secondary vehicle for the primary work [of women] in life: relationship maintenance.

One reader, Charles T. Galloway of Bolton, Ontario, compares mass restroom migration to earlier, seemingly anachronistic rituals:

> There are certain subjects that both males and females seem to regard as appropriate for discussion when the other sex is absent but unsuitable when they are present. The ladies' trip to the washroom takes the place of the old Victorian habit of the ladies withdrawing from the dining room to "leave the gentlemen to their cigars" so that these conversations can take place. What the ladies are doing in there so long is engaging in "girl talk."

David Ryback, a psychologist in Atlanta, Georgia, finds that "girl talk" is not only a way for women to increase intimacy but also serves to defuse what could turn into conflicts and turn them into ego-boosters:

> Women will often go to the restroom together precisely when they are in the company of men who take a dominant role in the discussion of the group. . . . The men are airing their opinions, controlling the direction of conversation and, for the most part, dismissing the women's contributions. The simple outlet for the building frustration on the women's part is to take leave of the men's company temporarily without challenging their pompous position. It may come at a time when one of the women truly needs to use the restroom. What better occasion for the other women to take leave of a frustrating, possibly boring, situation without the slightest hint of confrontation!
>
> Once in the restroom, the women now have an opportunity to regain their sense of self-esteem. They can do this by having a brief discussion of their own choosing. If they feel particularly put down by their male counterparts, they can restore the balance by sharing their opinions of the men awaiting their return. No wonder we men start fidgeting after a while. And then we wonder why the women look so radiant and self-satisfied when they return.

Nothing we heard from our readers contradicted Dr. Ryback's observations. Indeed, Karen Pierce of Springfield, Virginia, eerily echoed his words, and added, "I've had better discussions fixing my make-up than I've ever had talking to my boyfriends."

Men, on the other hand, aren't too likely to bond in restrooms; one reader, Bob Kowalski of Detroit, Michigan, thinks the reason might relate to Darwinesque theories:

> Men, probably harking back to their caveman hunter days, prefer to respond to nature's call alone. You don't want a prospective enemy too close at hand when you have your, uh, guard down!

Facilitating Conversations Outside of the Restroom

Readers brought up two points we had never considered. Diane Larson notes that since women seem to use the restroom much more often than men,

> it's better to have one lull in the conversation while several women go to the restroom at once. Otherwise, the table is constantly having to play catch-up because someone is always missing part of the conversation.

And Jennifer Talarico of Bethel Park, Pennsylvania, uses her group defections during double-dates to force the deserted males to bond:

> When we are in the restroom for extended amounts of time, the topic of conversation is undoubtedly the men we are with. And we know that the men back at the table are discussing us as well. Sometimes the purpose of our escape is so that the men can get to know one another, especially when we girls have been friends for a while and the guys have just recently met.

Conspiracy Theories

Several respondents have already suggested that the conversation regarding the abandoned males might be less than laudatory. This was very much on the minds of many of the males who wrote, including Wayland Kwock of Aiea, Hawaii:

As to why women go to the restroom in packs . . . it has to do with make-up. . . . The alternative, that they are making fun of their dates, is just unthinkable.

Sorry to do this, Wayland, but please meet Linda Lassman of Winnipeg, Manitoba:

. . . another reason women leave together is because they often think the things men say are *really* stupid, but they don't want to cause hurt feelings or arguments by saying so. Going to the restroom in groups allows them to talk about things that actually interest them, to discuss the same topics and have their opinions listened to, and to laugh at the men they're with without worrying about how the men will feel.

If we are indeed engaged in a war of the sexes, perhaps the paranoid musings of Ted Baxter might contain the true answer. Several readers, including Mark La Chance of Pleasanton, California, remember the "Mary Tyler Moore Show" news anchor's answer to this Frustable:

He believed that there was a women's plot to take over the world, and that their secret meetings were held in the ladies' rooms so that men wouldn't hear. I have to admit that after my history of disastrous and humiliating relationships, the idea of gender-guerrilla warfare rings true to me.

Sorry things haven't been working out, Mark, but by the time you finish this discussion, you will have delved as deeply into the female psyche as is genderly possible. Perhaps you will now be able to plumb the psyches of a woman as deftly as Richard Gere's *American Gigolo.*

And What Are They Doing in There for So Long?

Many of the explanations mimicked the discussions above. Several women cataloged the checklist of make-up and grooming that must be undertaken before any restroom exit. But more moaned about clothing problems. Diane Larson is typical:

I defy any man to don pantyhose, a girdle, a slip, a tight skirt and high heels and then go to the restroom in record time after squeezing

> into a tiny cubicle barely big enough to sit down in without your knees hitting the door.

And one type of apparel not to buy for women with weak bladders was mentioned by Joan Cartan-Hansen: "Woe to the woman who wears a one-piece jumpsuit and practically has to do a striptease to answer's nature's call."

Most of all, women wailed about the dearth of stalls in women's bathrooms. Nothing makes men prouder about their gender than cruising into a men's room at a ball game or concert while women stand glumly in long lines. Just in case men haven't noticed it, women wanted to point out that their anatomy is slightly different from men's. Even if women had more stalls in their restrooms and fewer make-up, grooming, clothing, and conversational distractions, it would still take longer for them to urinate than men. Several readers sent us graphic descriptions; Sharon Brandon was more discreet:

> Men have more plentiful "opportunities," shall we say, in their restroom, while women are often limited to a small number of closed, private stalls. I hope I don't have to go into grade school health class review to explain any other possible time-consuming differences to you between men and women.

No, thanks.

Another reader made it clear that even if a particular gaggle of women entering a restroom is childless, women with children can drastically affect restroom timing. Among the factors that Judy R. Reis of Bisbee, Arizona, cited in prolonging restroom visits are the following: "helping the kids go potty," "waiting for the women ahead of them to get done helping the kids go potty," and "relinquishing their places in line to the women whose kids can't wait to go potty." Obviously, some of the kids in the women's room are boys, not girls.

Most of the female respondents to this question were feeling a little sorry for themselves. But we know that all types of facilities are provided for women that are not given to men. At times, women's rooms look more like Ethan Allen showrooms, with all sorts of paraphernalia. Rosemarie Gee of Ridgefield, Washington, reminisced with us about the restrooms in the library at her alma mater, Brigham Young University. They provided couches and chairs, so a woman could eat lunch there, the only place besides a small room in the basement where one could sit, eat, and study at the same time. Some women's rooms even had beds! Rosemarie also mentions that many women's rooms contain full-length mirrors and chairs by the mirrors, to assist in undertaking all the tasks that make women take so long in there in the first place.

But even stripped of all fineries, Gee insists that a woman's task in the restroom is far more arduous than a man's, and she supplied us with a handy comparison chart to see how a man and a woman's trip to the restroom is likely to compare (your results might differ):

Men	*Women*
Open door or enter doorway	Open door or enter doorway
Step to urinal	Choose a stall
	Open door—maneuver in cramped quarters
	Hang up purse/coat
	Flush to ensure fresh water
	Wipe off seat
	Put on seat cover
Unzip	Unzip and pull down pants or lift up dress and pull down nylons.
Do business	Do business
	Use toilet paper
Flush (optional)	Flush
	Realign clothing
	Gather personal possessions
	Open door in cramped area
	Set personals on counter
Wash hands (optional)	Wash hands (likely)
Dry hands (conceivable)	Dry hands (optional)
	Gather stuff
Leave	Leave (eventually)

Obviously, if all the talking/make-up/child care/grooming behavior occurs as well, it is a wonder women ever emerge from the restroom at all. But faced with the thrilling prospect of rejoining their waiting specimens of male hunkitude, they always seem to come out eventually.

Submitted by Ray Bauschke of Winnipeg, Manitoba. Thanks also to Douglas Watkins, Jr., of Hayward, California; Edward T. Coglio of Pittsburgh, Pennsylvania; Ish Narula of Upper Darby, Pennsylvania; John Heggestad of Fairfax, Virginia; Bruce Kershner of Williamsville, New York; T. Wenzel of Charleston, West Virginia; and Alice Conway of Highwood, Illinois.

A complimentary book goes to Rosemarie Gee of Ridgefield, Washington. Thanks to the scores of readers who duplicated sentiments expressed above.

FRUSTABLE 2: ***Why do men tend to hog remote controls and switch channels on television sets and radios much more than women?***

Right after the publication of *When Did Wild Poodles Roam the Earth?*, *Consumer Reports* published the findings of a survey conducted among their readers about remote control usage (thanks to one of our favorite correspondents, Kenneth Giesbers of Seattle, Washington, who called this to our attention). Their study indicated that men are more than twice as likely to hog the TV remote (38 percent to 15 percent). And not only do men "channel surf" more often than females (85 percent to 60 percent); they are less likely to complain about their mate's surfing (66 percent to 43 percent).

Although the Frustable, as posed, refers to channel switching on radios as well, we received little response to radio station hopping. In fact, Jennifer Talarico, of Bethel Park, Pennsylvania, while concurring that men switch car radio channels far more often than women, did not agree that they exhibited this behavior at home. What accounts for the discrepancy? "At home, a man is too busy watching all three hundred channels on television to be preoccupied with the radio."

How can we account for this male obsession? More than a few folks had a simple explanation, most eloquently stated by Donald Wiese of Anaheim, California: "Maybe men are simply jerks!"

Certainly a plausible theory, Donald. Indeed, most of the conjectures were not ones that would deepen men's self-esteem. About the sunniest possible explanation that we received was that men are obsessed with gadgets and will play with them regardless of whether it advances any particular goal. Most correspondents, though, found far darker reasons for hogging remotes:

1. Men Need to Dominate and Control

"The remote control gives the man power. Plain and simple," responds Kelli Zimmerman of Milwaukee, Wisconsin. Lauren Goldfarb of Huntington, New York, issues a call to arms:

> The male gender tends to dominate more than the female. Like it or not, the reality is that this world is still run by men. So you ladies at home, take control and refuse to give up the remote. It may be well worth it in the end.

So the battle of the sexes is waged not in boardrooms but in living rooms. Concurring with this sentiment is Kathy Smith of New Bern, North Carolina:

> My husband is a "remote controlaholic" because he can't control *me*—it's a displacement behavior that gives him a feeling of accomplishment and superiority.

2. Men Are Hunters

Jerry Seinfeld has postulated that channel surfing is a modern equivalent of hunting for men. Lauren Goldfarb concurs:

> In the cave days, men were hunters while women nurtured the family. The action of flipping the channels on the remote control is similar to the hunt. A man with a remote control in hand is a man with power, hunting for something exciting and interesting.

3. Men Require Instant Gratification

"Perform on command or I'm off," says David Ohde of Weaverville, California, is the watchword of most men. Lane Chaffin of Temple, Texas, adds that because men watch so many sporting events, where dead time is clearly demarcated, this tendency is exaggerated.

4. Men Are Promiscuous

Why do men require instant gratification? Because they are used to insisting upon control/dominance, according to the readers mentioned above. But Lauren Goldfarb is back with a theory on this subject, too:

> Most men don't like to commit or get attached to just one show, which is not all that different from a typical teenage boy. A girl dreams of her wedding day while a guy dreams of how many women he will "have." Why just have one when you can have them all?

Several readers indicated that women are much more willing to commit, in time and emotion, to one program. Who would have thought that remote control hogging could be directly traced to a fear of intimacy?

5. Men Are Mice

Men are animals, insists Karen Flanery, of Casper, Wyoming, and act like any other creature that scientists have investigated:

> Remote controls are a prime illustration of the response-reward theory advocated by early psychiatrists. Mice, simians, canines, and felines soon learn to press a lever that will give them a reward (say, a piece of cheese). If the response is intermittently rewarded, the drive to press buttons intensifies.

Men didn't receive much sympathy from our female readers, but we did receive a poignant note from Ruth M. Johnson of Tacoma, Washington, which testifies to the primal connection among males, channel surfing, and the beloved remote control:

This is purely a matter of control. The remote control device is an ideal way to drive a female out of her mind.

My husband died of Lou Gehrig's disease in 1991 and the remote control was the last thing he was able to operate. At the end, he had to have a holder on his palm, which held a pencil to enable him to punch the buttons, but he never failed to change the station the minute I became engrossed in a program or to skim the channels so fast my eyes would glaze.

It was the only thing he could do for himself, so I let him carry on. The television has been on exactly three times since he left me for a better place.

Submitted by Patricia M. Delehanty of Poughkeepsie, New York.

A complimentary book goes to Lauren Goldfarb of Huntington, New York.

FRUSTABLE 3: *Why do some women kick their legs up when kissing?*

Some of you believed that leg-kicking kissers are merely imitating the lovers in romantic, old movies. But this begs the question. Then why did the heroines in old movies kick their legs up?

Most of you were prosaic. You thought it had to do with a simple, anatomical truism: Women tend to be shorter than men. Bob Kowalski of Detroit, Michigan, had a typical response;

> Maybe women kicking a leg up while kissing has something to do with most women having to reach up to kiss their love! Most men, being taller, kiss down. The leg up may be an automatic balancing response, and they are probably unaware that they even do it.

Of course, kicking doesn't bring them straight up; it also moves women forward, as Kelli Zimmerman explains:

> Women stand on the tiptoe of one leg while lifting the other so that they can lean forward. Why we don't just ask the tall men to bend down a little bit is beyond me.

Maybe, Kelli, the reason is that the leg kicking facilitates more than just lip clinching, as David Ryback postulates:

Women . . . can get their upper bodies closer to the men of their desire by standing on tiptoe and kicking up one leg behind them. If you find this hard to believe, try standing close to a wall. Then lift one leg behind you. You'll find yourself "hugging" the wall.

We just did. Thanks, David, that's the closest we've come to an intimate relationship in months.

Although all of these explanations make some sense, we have a nagging suspicion there is more to the issue. Lane Chaffin of Temple, Texas, is the only reader who indicated that height is not the only factor in leg kicking. He has never seen a short man kick up his leg when kissing a taller woman,

> probably because it would make the man feel weak and the (taller) woman would feel uncomfortable because she would be in a physically awkward position. Maybe the leg kicking started as a woman's exhibition of trust in her male companion.

Could be. Our informal survey indicates that leg kicking occurs only in public places (a classic place: airports, when loved ones greet one another). And leg kicking is clearly a romantic gesture: If it were merely a convenience to compensate for height differences, why don't daughters kick up their legs when kissing their fathers good night? Or little kids when kissing their taller grandparents?

We wouldn't be surprised if the reason why we find leg kicking in public places is that the gesture is used by the woman, usually subconsciously, as a marker, to stake a claim that "this man is mine"—to the world and to other women in particular. Outlandish? Not really. When we see couples strolling down the street, with the man putting his arm over the shoulder of his companion, it is a way of telling the world, "She's taken." Yet in the privacy of their own home, husbands and wives seldom walk hand in hand from the dining room to the kitchen to do the dishes after a meal or while taking the garbage outside.

Anyone have any better theories?

Submitted by Jerrod Larson of South Bend, Indiana. Thanks also to Rosemary Lambert of Kanata, Ontario.

FRUSTABLE 4: ***Women generally possess more body fat than men? So why do women tend to feel colder than men in the same environment?***

Many readers tackled this Frustable, but we were most impressed with the arguments of two professors, especially because their two discussions complement each other. Dallas Brozik, chair of the department of finance and business law at Marshall University, theorizes why the extra fat might make a woman feel cooler rather than hotter than men in the same room:

> The reason that women may feel chilled even with an extra layer of body fat has to do with the body's attempt to maintain a central core temperature. As the body tries to maintain 98.6 degrees Fahrenheit, the blood system is used to transfer inner heat to the skin, where it can be radiated away. Heat transfer can only be accomplished through a temperature gradient across a boundary, and the additional layer of fat makes it more difficult for women to rid themselves of excess heat.
>
> At this point the body has two mechanisms it can employ to rid itself of the excess heat. First, it could raise the core temperature to establish the proper temperature gradient with the ambient external temperature. But the body is trying *not* to heat up, so this mechanism is self-defeating. The second mechanism is to sweat so as to bring evaporative cooling into play. The extra layer of fat makes women sweat a little more than men under the same conditions. And when this sweat is evaporated, the nerve cells in the skin feel the chilling effect; hence, women will tend to feel colder than men under similar conditions.

But other physiological forces are at work, deftly explained by Richard Landesman, of the department of zoology at the University of Vermont:

> Body temperature is the result of metabolism, and, at rest, the bulk of the heat to warm the body is produced by the liver, heart, brain, endocrine organs, and skeletal muscle. The latter is responsible for about 30 percent of heat production at rest. During exercise, the heat produced by the muscles contributes significantly to body temperature.
>
> There are two temperature regions of the body: the core, whose temperature remains relatively constant; and the shell or surface, whose temperature tends to vary with physical and environmental changes.

There are many mechanisms to raise and lower the temperature of the body: for example, shivering raises the temperature and sweating lowers the temperature. Another way to conserve heat is for the blood vessels in the skin to vasoconstrict, thereby shunting the warm blood to the core of the body. One obvious symptom of vasoconstriction is for the skin to feel cold. Now with that preamble, the answer to the Frustable . . .

> 1. As a general rule, men have more muscle mass than do women; therefore, men can maintain their body temperature at rest without feeling as cold as women.
>
> 2. Women do have more subcutaneous fat compared to men. This layer serves to give the women body shape as well as to provide a layer of insulation. When it is cool, the blood vessels in the skin vasoconstrict, shunting the warm blood into the core of the body. The skin now feels cool. The layer of insulating fat, while conserving the heat in the core of the body, contributes to the skin remaining cool.

Submitted by David Held of Somerset, New Jersey. Thanks also to Bruce Kershner of Williamsville, New York.

Complimentary books go to Dallas Brozik of Huntington, West Virginia, and Richard Landesman of Burlington, Vermont.

FRUSTABLE 5: *Why is the average woman a much better dancer than the average man?*

Fred Astaire, Mikhail Baryshnikov, and Gregory Hines aren't exactly chopped liver in the dance department, so we know that men *can* dance well. The conundrum is why the average man doesn't.

Our readers came up with five possible explanations:

1. Girls Practice Dancing More Than Boys

This was by far the most popular theory among *Imponderables* readers. Jody Jamieson Dobbs's response was typical:

Women start at an early age dancing with their moms, sisters, and friends—they don't need male partners. Men wouldn't be caught dead dancing with each other. So females get to practice from a very early age and, let's face it, practice makes perfect. (I'm an excellent dancer and my husband is an excellent hunter and fisherman.)

Ah, the typical American family: the wife is graceful, while the husband is proficient at protein-gathering.

But why don't boys dance together . . .

2. Dancing Is for Sissies

Western culture has deeply conflicting feelings about male dancers. On the one hand, fictional depictions of ballroom dancers, ranging from Fred Astaire to John Travolta to Gene Kelly to Patrick Swayze, invariably portray the male as virile and extremely attractive to women. Yet the general attitude of the average boy has not changed much since Robert Coulson of Hartford City, Indiana, was avoiding the dance floor:

> I can't speak for the current generation, but when I was a boy, dancing was considered "sissy." I went to a rural school, and I don't think any male ever learned to dance while I was there. From TV reports, boys today seem willing to learn the more athletic dances, such as break-dancing, but I suspect that ballroom dancing is still considered beneath contempt.

Curiously, though, dancing has often been considered a macho pursuit among diverse working-class subcultures, ranging from the Italians depicted in *Saturday Night Fever*, the spiritual descendants of the kids who danced on "American Bandstand" in the 1950s, to the African-Americans in inner cities, where break-dancing and hip-hop cultures emerged.

Another correspondent, Joseph M. Novak of Pittsburgh, Pennsylvania, argues that from the earliest ages, boys are touted toward what society deems gender-specific physical activities emphasizing strength and power, such as football and wrestling. Girls, on the other hand, are encouraged to develop grace and beauty, in sports such as gymnastics and arts such as ballet. Novak even argues that the musculature of the two genders might play a factor: "Who do you think is a better dancer, Arnold [Schwarzenegger] or Maria [Shriver]?"

We're betting on Maria. But we're still concerned that we haven't pinned down why dancing is considered a feminine form. Some readers feel the answer is . . .

3. Dancing Is an Emotional Expression, and Men Don't Like to Express Their Emotions

Cheryl Stevens's response speaks for several readers:

> Dance is an emotional expression, like any art. Men are not (or were not) encouraged to express their emotions, especially with potential spectators around. Maybe as men are allowed to be more sensitive, they'll become better dancers.

Cheryl, we're betting that big bands will come back before men become more sensitive. But we like your point about spectators. Men tend to open up emotionally to a few selected intimate friends, whereas women are more likely to share emotions with acquaintances; therefore, men might be more reluctant to express their emotions (and thus their vulnerability) in public, even through dance.

Still, why would men have no problem expressing emotion in writing or music but stumble upon dance? Perhaps the answer is that while we don't tend to read the average Joe's prose or listen to an amateur's musical stabs, we are constantly subjected, whether at wedding parties or at discos, to the prancing of amateur dancers, many of whom are coerced onto dance floors by their dates or mates. After all, there are plenty of spectacularly graceful *professional* male dancers; maybe the average man would be just as reticent about playing the oboe or reciting his poetry in public.

4. Men Don't Like to Dance, So They Don't Do It Well

This may not be the most profound explanation, but it makes some sense. For all the reasons stated above, many men don't like to dance. LaNue Parnell-Reynolds of Warren, Arkansas, thinks this insight unlocks the Frustable:

> You need to watch some *country* dancing! Men are good at what they enjoy. More women than men like ballroom-type dancing.

5. Dancing Seduces Men More Easily than Women

We have no idea if Robin Pearce's theory is correct, but it was the only answer we received that ties together the many threads we've discussed in a way that, well, might bag her an "A" in an anthropology course:

> Dancing, like brightly colored clothing and makeup, is a way of visually displaying one's self to potential mates. Since men are visually oriented, women take greater pains to make themselves attractive [to men], and this includes dancing well; gay men are usually excellent dancers for the same reason [they are also trying to attract men].
>
> Your average straight white American male subconsciously feels that dancing well, like being too concerned with one's appearance, is not masculine. However, heterosexual men in cultures where visual display is considered appropriate for the male (i.e., Latin, Mediterranean, and black men) both dance well and dress colorfully.

Robin is obviously generalizing about gender and various ethnic and national types, but her notion is fascinating. After all, choreographers and lyricists constantly cast dance as a metaphor for sex. And although it is hard to conjure up any woman being enticed by the fellow performing a lumbering lambada at the disco, it is hard to argue with Pearce that the ultimate, if often subconscious, goal of much social dancing is seduction.

While most men are attracted *by* the dancing of potential mates, it doesn't necessarily follow that most men are attractive *when they dance.* Maybe the maladroit average male is showing rare good sense in being a wallflower—better not to attract a mate than to lose any chance by flailing around on the dance floor.

Submitted by Judith Dahlman of New York, New York.

A complimentary copy goes to Robin Pearce of Kansas City, Missouri.

FRUSTABLE 6: ***Why do so many people put their hands up to their chins in photographs?***

"Because the photographer tells them to!" answers Rosemary Gee of Ridgefield, Washington. She's right, in a sense. We doubt if many portrait subjects spontaneously thrust their hands into their chins upon hearing "cheese." As Bill Jelen of Akron, Ohio puts it:

> It is a conspiracy by all of the professional photographers. Have you ever seen anyone posed this way in an amateur photograph? No!

We can testify to the existence of hand-in-face commands by professional photographers. We didn't spontaneously place our fingers on our forehead when the picture on the back cover was shot. We were asked to do so.

Still, we need to plumb deeper. Even if only professional photographs seem to feature this pose, why do these photographers request it? We spoke to many professional photographers and received quite a bit of response from readers, too. As usual, with Frustables, answers were all over the map; but they fell into five major categories:

1. Look Ma, No or Two Hands

We received this unusual theory from James P. Gage of Washington Island, Wisconsin:

> Many portraits were requested by immigrants to send to relatives in the "old country." Should the fingers not be shown, the relatives would assume the immigrant had lost some or all of the digits in transit or at work. Even in the 1960s, when I worked as an apprentice at a portrait studio, "Show their fingers!" was the request of the boss.

Finger-posing can exhibit good news as well as bad. Gail Lee Dunson of Dallas, Texas, points out that "in a wedding or graduation portrait, the pose is used so that the 'portraitee' can show off the new ring."

2. A Cheap Form of Plastic Surgery

The most popular theory among readers is that the pose is used to cover a myriad of physical defects in the subject: wrinkles in the neck; a weak chin; a double chin; acne or other blemishes; a scar; or even bad teeth.

Of course, no subject is going to look great if the image is blurred. Curtis Krause of Vernon, Connecticut, offers a theory similar to that mentioned earlier in this book during our discussion of why people didn't smile in old photographs:

> Many old photographs contain blurs caused by movement during exposure. Exposures were originally rather lengthy. . . . It may be that the hands on the chins helped to steady the head and avoid blurring. As film speeds increased, this pose would have become less necessary, but might have been continued as part of the style.

Ultimately, the more important reason why the hand-to-chin pose might make the subject look better was expressed by John C. White of El Paso, Texas:

> People posing for photo sessions often feel awkward and don't know what to do with their hands. Photographers have learned to resolve this problem with the hand-to-chin pose. It may be a bit hackneyed but it sure beats the Napoleonic pose.

3. Spruces Up the Image

Many readers ("Dobie Gillis" fans?) mentioned that the hand-in-chin pose reminds them of Rodin's *The Thinker.* To them, the pose conjures in the viewer the notion that the subject is an intellectual, of a philosophical and cerebral nature.

Gene Lester of the American Society of Camera Collectors notes that hands can enhance the subjects in other ways:

> Hands are very expressive and the way people hold them or use them shows a bit more of their character. Hands can help express deep thought, puzzlement, even humor, depending upon how the subject uses them.

Reader Joanna Parker of Miami, Florida, corroborates and expands upon Lester's thesis. If you want to show the potential power of this pose, suggests Parker, watch actors:

If you watch truly adept actors, you will find that they consider the body to be three spheres, balanced one atop the other. When you want to express power, you use all your arm and hand gestures in the middle sphere. You punch from there, poke your fingers into the other guy's chest from there, tell people "no!" with a slicing gesture of the open hand from there, etc.

If the actor wants to exude sex, he works with his arms/hands in the lower sphere. All gestures originate from and work in this area.

The top sphere projects intelligence, thoughtfulness, and intimacy. Watch good actors who are playing professors. Chances are that when they are on camera, they are fiddling with their glasses, moving their hair off of their foreheads, using a pipe by gesturing with it held close to their faces, etc.

In still photography, perhaps the placing of the hands under the chin enhances this expression of intimacy with the viewer of the photo. It lends the photo a feeling of sincerity.

4. Improves the Composition

Several readers and professional photographers noted that a head floating in space looks funny in a photograph. The hand-in-chin pose anchors the head, but it also makes the composition look more natural, at least according to photographer Mike Brint, who notes that in nature, larger objects usually support smaller ones. He speculates that the originator of the pose might have thought that a head shot featuring a thin neck supporting the head looked funny, like an ice cream cone.

Why might the hand-in-chin pose create a superior composition? Photographer Bill O'Donnell supplies an explanation:

> The most common explanation is that the vertical or diagonal line formed by the arm will tend to "lead" the viewer's eye to the face. A simple face-front picture looks rather static and may be unflattering; the hand-in-chin pose makes the subject appear more relaxed and life-like. There's a reason why mug shots and driver's license photos look like they do—no hands, no relaxed pose, no suggestion of possible movement.

Chandra L. Morgan-Henley of Cleveland, Ohio, is also a proponent of the "composition theory" and notes that good photographers

> compose photographs in their minds before clicking the shutter. . . . What might otherwise be a poorly composed and boring portrait can, with the addition of a carefully placed hand, strike exactly the right balance for visual interest.
>
> Please note that I said *can* in the above sentence, because poorly trained (and untrained) photographers often copy techniques that they have seen, with results that are less than pleasing. Incidentally, my father and two brothers are all portrait photographers, and I studied retouching before deciding my talents lay elsewhere—such as in writing letters to authors.

Good idea, Chandra, there's a great future in the lucrative and expanding field of writing letters to authors.

5. It Sells, Stupid

The chin-in-hand pose was met mostly by derision by the professional photographers we contacted. Yet many noted that the pose is consistently popular with subjects. And if subjects want a pose, even if photographers disdain it? In this case, presumably, the customer is always right.

Typical, if slightly more emotional than most, was the reaction of professional Wilton Wong of Belmont, California:

> I get nauseated every time I encounter a senior photo on someone's wall or desk, with the senior shown with the hand up to the chin or side of head. It is obviously so phony, since one can tell that absolutely no weight of the head is resting on the hand (since that would cause facial distortion). . . . As for why this is done, I'll speculate that it is merely the result of a photographer who shoots "by formula" as learned in a workshop ("Now remember, pose #147 is great for seniors!") and relies on cliche poses rather than being a real artist and coming up with innovative ways of making the photo interesting.

Over the last five years, we've harassed Joann Carney, the talented photographer who managed to keep the *Imponderables* author photo down to only a minor sales deterrent. "Why did you suggest the hand-on-forehead pose for this photo?" we prodded.

Believe it or not, Carney didn't know herself. She mentioned that amateur subjects are often nervous and that movement distracts them from focusing on the mechanics of the shoot; she noted that some people come alive when using their hands; and she muttered something about the composition being more interesting. Maybe, in a burst of psychic energy rarely encountered, she anticipated the appropriateness of this pose for the title of this book.

Submitted by Alice Conway of Highwood, Illinois.

A complimentary book goes to Joanna Parker of Miami, Florida.

FRUSTABLE 7: ***Why do very few restaurants serve celery with mixed green salads?***

It's green. It's crunchy. It's low in calories. It's in tuna and chicken salad. Can you think of another vegetable, besides lettuce, more qualified to belong in a mixed green salad than celery?

But a glance at just about any mixed green salad will reveal a lack of celery. What's going on? Mark Watson of Cary, North Carolina, formerly responsible for the preparation of the salad bar at the Chapel Hill Country Club restaurant, weighs in with three important reasons:

> First, every competent restaurateur knows that "people eat with their eyes." Many of the ingredients in any item served in a restaurant are there as much for visual appeal as for flavor. Since celery is the same color as the lettuce base, it fails to improve visual appeal. On the other hand, tomatoes, carrots, and onions (white or purple) provide contrast. I cannot overemphasize the importance of visual presentation in the restaurant business. I have seen time and again that a good meal, poorly presented, will generate complaints, while a mediocre meal, presented colorfully and artistically, will bring raving compliments.

Second, the preparation of celery is relatively labor-intensive. It must be washed, and the widely flared white base must be cut off because it tends to be too fibrous and ugly. The leafy tops and often the narrow branches are removed, and then any damage spots on the stalk are excised. Finally, it is usually chopped into fairly narrow slices.

Third, the one superior attribute that celery does have disappears quickly in storage: that crispy crunchiness. Another maxim in the food industry is that food should have "good mouth feel." Celery provides a satisfying crunch that consumers (correctly) associate with freshness. Leftover chopped celery tends to turn brown and become rubbery rather quickly compared to other vegetables; this increases food waste. Restaurants used to solve this problem by storing celery (and lettuce and sliced potatoes) in a container of water mixed with a powdered vegetable preservation product that contained sulfites. To the best of my knowledge, this process has been widely abandoned, because the FDA discovered an annoying drawback to the practice. People all over the country were eating sulfite-treated salad vegetables, keeling over, and plummeting into life-threatening allergic anaphylactic shock!

Ronald J. Moore of Ecorse, Michigan, a professional chef for more than thirty years, backs up Watson's complaint about the quick spoilage of celery. Restaurant personnel don't appreciate having to prepare celery for dinner just hours before customers enter:

> If salad is kept from, say, 2:00 P.M. to 9:00 P.M., it is fine without celery. If it contains chopped celery, seven hours will have turned it from mixed green salad into mixed brown garbage.

Even before it oxidizes, celery tends to be dirty. Travel writer Judy Colbert reports that her food service friends tell her that celery is full of bacteria and is particularly difficult to clean properly. Another complaint that Judy and Ronald Moore mention is that celery is stringy and tends to get caught between teeth. Moore reports that older customers complain that celery gets trapped in their dentures.

Sure, celery adds crunchiness to a salad, but reader Cheryl Stevens notes that the cabbage and lettuce in a mixed green salad also provide snap. Tuna and chicken salad, on the other hand, "squishy in consistency," require celery for texture.

Finally, Ronald Moore contributes what could be a decisive factor in the demand-side of the missing celery equation: "An amazing number of people absolutely hate the taste of celery."

Submitted by Malcolm Boreham of Staten Island, New York. Thanks also to Launie Rountry of Brockton, Massachusetts.

A complimentary book goes to Mark Watson of Cary, North Carolina.

FRUSTABLE 8: ***In English spelling, why does "i" come before "e" except after "c"?***

We're afraid that as of now this Frustable is frustrating our readers as well as us. But we have just begun to fight.

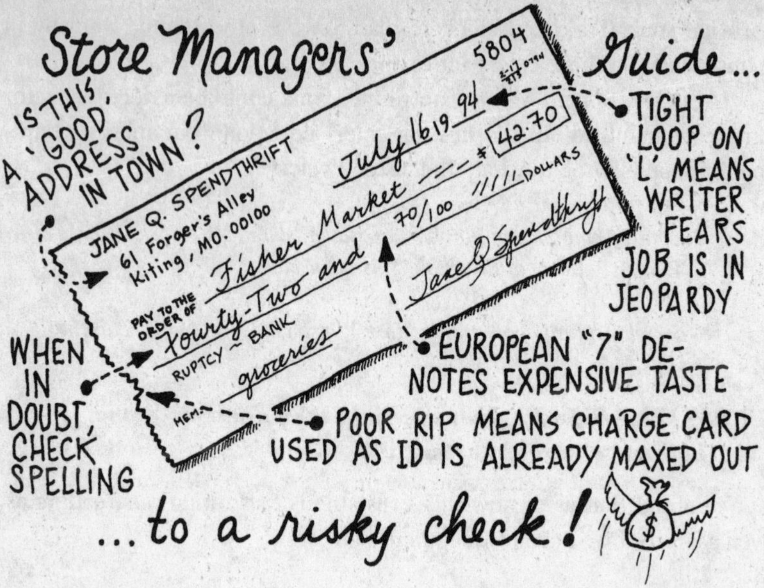
Store Managers' Guide...
IS THIS A 'GOOD' ADDRESS IN TOWN?
JANE Q. SPENDTHRIFT
61 Forger's Alley
Kiting, MO. 00100
5804
July 16 19 94
PAY TO THE ORDER OF Fisher Market
$ 42.70
Fourty-Two and 70/100 DOLLARS
RUPTCY ~ BANK
MEMO groceries
Jane Q Spendthrift
TIGHT LOOP ON 'L' MEANS WRITER FEARS JOB IS IN JEOPARDY
WHEN IN DOUBT, CHECK SPELLING
EUROPEAN "7" DENOTES EXPENSIVE TASTE
POOR RIP MEANS CHARGE CARD USED AS ID IS ALREADY MAXED OUT
...to a risky check!
$

FRUSTABLE 9: ***What in the world are grocery store managers looking for when they approve personal checks?***

When we first posed this Frustable, we wrote:

> We have been most dissatisfied with the answers we've received from supermarket chains on this topic, so we're hoping that some grocery store checkers, managers, or perspicacious customers can help us with this Frustable. To us, it seems that the manager simply peeks at the check, glances at the customer, and approves the check without looking for anything in particular. In fact, we've never seen a check rejected.

We've never dealt with a question pertaining to a particular profession that has generated so much mail. If supermarket chains were reluctant to discuss this issue with us, current and former grocery store checkers and managers were not reticent at all. We received dozens of letters from readers in the trade, including a magnum opus from Stephen H. Cook of North Providence, Rhode Island, who has spent twenty-three years in the personal credit and retail computer systems business and has done research on the machines that provide electronic approval of personal checks.

Cook confirms that grocery store managers are often negligent in their duties but notes that they must be on the lookout for four types of bad checks:

> 1) bad checks deliberately written by persons who know it's illegal
>
> 2) checks written by persons who don't know how to do a good job of using a checking account (the careless and disorganized)
>
> 3) bad checks from people without overdraft protection on their checking account, who know they can't cover the check at the moment but are hoping that a recent or planned future deposit will clear before the check does
>
> 4) bad checks from those who have made a recent deposit but have erred in assuming when the check will clear (these folks also have no overdraft protection)

According to Cook,

> Most bad checks fall into categories two and three. All checks in category four and some checks in categories two and three may be made good by being presented again to the bank. But checks in category one are totally worthless and represent the biggest risk to the retailer.

Although the store fears the bad-check artist most, some of the problem items that a manager scans a check for are just as likely to arise not only with the other three types but with customers whose checks are valid. For the protection of the store, managers may look for all of these red flags:

- Is the check signed?
- Does the written amount on the check match the numerical amount?
- Are the address and phone number of the customer printed on the check? (Many grocery stores will not accept starter checks or checks without imprinted addresses; at the very least, they will insist upon other identification, such as a driver's license or credit card, confirming the address of the shopper. In this case, a credit card not only corroborates identification but serves as a credit check.)

• Is the customer on a bad check list published by a consumer credit clearinghouse or by the grocery store (chain) itself?

• Does the name signed match the name imprinted? Former store manager John Schaninger of Easton, Pennsylvania, reports, "I once had an assistant manager who did not check this; when I received the bad check back from the bank, the signature read, "I Beat You."

• Is the check made out to my store?

• Has the check been tampered with? Per C. Clarke, night manager of a Hy-Vee Food Store in Spencer, Iowa, reports he recently saw a check with a "void" clearly etched but (mostly) erased.

• Is the date correct? Let's use this seemingly innocent item as an example. Joseph S. Blake, Jr., a store manager, wrote to us about all of the potential problems inherent in an incorrectly dated check:

> If a check is postdated, it is legally considered a promissory note and not a check. If it is returned, the person who wrote it merely has to say, "I asked him if he could hold it until the date on the check and he said it would be okay." Although I would never agree to such a condition, the bank endorsements would clearly show that I deposited it before the date on the front. I would be screwed.
>
> I note that the year is correct, particularly during the first couple of months of the new year. A check that is dated more than six months ago is considered "stale dated" and cannot be collected as a cash item. . . .
>
> While I may have just thought that it was a confused little old lady that inadvertently wrote May 10 when it was really April 10, I find out later that she has done the same thing all over seven states. She has thousands of dollars of judgments against her, but no one can ever collect, since her only real asset is her checkbook for an account that never has more than five dollars in it.

All of these eight red flags are cut and dried; there is no particular reason why a checker wouldn't be qualified to verify them. But some verifications require a little more skill to examine. Blake named the three most important:

• Is it a genuine ID? Anyone can obtain a fake ID for a nominal amount. Many of them are horribly executed, but some can fool an inexperienced checker.

• Does the photo ID or description on a driver's license match the person trying to cash the check? As Blake puts it,

> If someone were to break into your car or home, he could easily end up with your checkbook as well as several pieces of your identification. It would be less likely that the ID would match your height, weight, hair color, and other features. If the check were stolen from someone, my only recourse is against the person who wrote the check, not the person who had it stolen from him.

• Is the check writer signing the check in the manager's presence? Some managers insist on seeing the customer sign the check in front of them (in our experience, this is not the case), and Blake offers a reason we would have never thought of, naifs that we are:

> Here's a popular scam from a few years ago that still happens from time to time. Suppose that you and I were in the business of writing bad checks together. Before I entered the store, I would have you sign my name of one of my checks. I would then go into the store with my checkbook with the check that you had signed, pick up all the big-ticket items that I could load in my cart, then proceed to the checkout counter. I would have my own identification, and my own checkbook where you had signed my name on the check, and would complete the sale.
>
> What could be more perfect? Unless the clerk or manager noticed that the signature on my ID and the check were not the same? We could do this in a couple of dozen places, preferably on a weekend. I would then wait and call the bank on Monday morning and notify them that my checkbook and wallet had been lost or stolen, or even wait until I started to receive overdraft protection before I "realized" it.
>
> An examination of the signatures would reveal that they were not mine—and they wouldn't be. The checks would all be returned to wherever they came from, and the merchants would be stuck. This is the reason why many places installed cameras that take a picture of the person writing the check.

Obviously, a bright and inexperienced cashier could be trained to perform all of these check-verification functions now usually undertaken by the store manager. We understand why supermarkets might want to lay responsibility on more experienced personnel, but isn't the practice a horrendous waste of time? Maybe there isn't a logical reason why the manager has to approve each check personally. Certainly, Russell Shaw, a journalist who writes frequently for *Supermarket News*, the largest publication covering supermarket management, has a jaundiced view of the check approval process:

> There's an insidious reason why supervisors approve checks, and this is where corporate culture fits in. Unlike some more progressive and newer industries, supermarkets have never been known as places to empower their hourly workers (read, cashiers). If you listen to a cashier paging a manager on the PA system, the manager will always be referred to by a surname ("Mr. Smith, register 8"), even though supervisor Smith's $18,000 a year salary and high-school education might pale next to the $100,000 salary and MBA of an executive at a nearby office park where the secretaries call him by his first name.
>
> Some of these attitudes are due to a caste system driven by low pay and low skill levels fostered by the tight profit margins of supermarkets. Many hire young and inexperienced cashiers, and management doesn't trust seventeen-year-old cashiers.
>
> As this feudal system affects the supervisors' relations with cashiers, it affects relations with their supervisors as well. In many cases, supervisors approve checks because there is little else for them to do. Virtually all the product-buying decisions are made at the regional or district-manager level. In some supermarkets, supervisors have little responsibility other than scheduling of workers. This gives them something else to do.
>
> It's a fact, Dave, that when a line is held up while a supervisor is paged for check approval, customers get annoyed. Yet this fact seems to be ignored by the by-the-book types who write thick policy manuals yet are oblivious to customer convenience.

This inconvenience is what prompted this Frustable in the first place. If customers are going to be subjected to a long delay while the manager ambles over to the checkout line, at least the inspection of the check shouldn't be so cursory.

Of course, the upper management of supermarket chains could argue that the identification amassed by a supervisor can be vitally important. Lawyer Jim Wright of Decatur, Georgia, informs us that the gathering of backup verification (e.g., social security number, driver's license number) grants the grocer immunity from civil or criminal liability for false arrest or malicious prosecution if the store prosecutes someone:

> In short, by obtaining verification, the manager is covering his butt should the check be bad and a warrant taken out. Quite frequently, when someone is arrested for writing a bad check, the accused claims that some other family member wrote the check without permission. The checking account holder than countersues the retailer for false arrest. If the verification of identity is written on the face of the check, the grocer is granted immunity from all liability for bringing the bad check warrant. He's not looking to reject your check; he's looking to protect himself.

A fascinating thesis, but one that wasn't mentioned by a single retailer.

In fact, we must confess that we have personally approached many grocery store cashiers and managers about this topic. When we ask what they are looking for when they take a passing glimpse at ours and others' checks, the answer is usually, "We're looking at the check number." Typical was the response of Fran Burns of Moneta, Virginia:

> Many banks print the month and year the account was opened on the check. If that isn't printed, we take note of the check number. Presumably, a low check number is a newly opened account. On a recently opened account, we generally will do some additional checkup on the customer.

The "low-number" theory was by far the most popular explanation from our readers, as well. We heard from some employees whose stores use 150 (or 49 checks past the usual 101 first number of checks), 300, and 500 as the demarcation line for additional security, whether it is more identification, calling the bank, or, in some cases, refusing to cash low-numbered checks or starting checks. Still, as we have discussed before, most banks allow new customers to start checking accounts with any number they want on their first checks (many businesses don't want to broadcast the fact that they are starting a new enterprise, and we'd be surprised if cunning bad-check passers would either).

Stephen H. Cook chronicles a cagier strategy:

> Most bad-check crooks have to move fast for their schemes to work consistently. This forces them to use checks with low numbers. The first rule for retail store managers is to be suspicious of low check numbers. Managers are supposed to engage in conversation with anyone presenting such a check, starting with questions along the line of, "New in town? Tell me about your apartment or house? Have you met your new neighbors yet?"
>
> The grocery store manager is then supposed to point where the cleaning and related household supplies aisle is. If the shopper is legitimately new in town and is unfamiliar with the store, this apparent friendliness on the part of the store manager is very welcome, and is good business practice for the retailer. If the person is a crook, all of this attention from the store manager should force a decision to take the bad checks elsewhere.

Sounds like Cook is implying that every grocery store manager is a would-be Columbo, making criminals squirm. And several of our readers were sure that this thesis explains why the examination of the proffered checks is often so cursory. Lane Chaffin of Temple, Texas, wrote:

> The whole rigmarole is to make a would-be bad check writer "sweat it out." If the check writer feels he may be turned down or possibly even recognized as a previous violator, he may change his mind.

Cook confirms that part of a manager's job is to "size up" check writers, and that certain profiles persist that can lead to prejudicial judgments. According to Cook, the best risks for personal checks include:

> 1) clean-cut parents shopping with children
>
> 2) clean-cut father shopping with children
>
> 3) well-dressed, well-groomed woman (especially if shopping in the evening after work and before dinner)
>
> 4) male coming in after work, still wearing a necktie, who is buying a few items
>
> 5) any clean-cut person who is making a large purchase and who is able to present enough cash to cover most of it, but needs to write a check only for the discrepancy

According to Cook, these are the worst risks:

1) single males buying staples

2) harried, "thirty-something" women, especially if overweight, unattractive, or shopping with children

3) married males buying staples, who present little cash and no credit cards

4) any shopper using food stamps

Says Burns, this list represents the following prejudices:

1) Single males are notorious for being sloppy and disorganized about household habits, including the proper maintenance of a checking account.

2) Unattractive divorced women with children have trouble making ends meet.

3) Suspicion of any married man doing the family grocery shopping by himself.

4) Any person using food stamps is too poor to properly maintain a good checking account.

Some of these prejudices may have enough basis in fact to perpetuate themselves, some are just based on outmoded traditions about who does the grocery shopping.

Grocery store managers are supposed to be suspicious of personal checks because grocery shopping has traditionally been a cash-only business. As a result, grocery stores get more bad checks than other types of retailers from people who are *not* crooks.

In conclusion, the answer to our Frustable is: Grocery store managers can be looking at about twenty-five different things when they give your check the once-over. But if they approve it as quickly and as superficially as most seem to do, either they are not doing their job right or you live in Mayberry R.F.D., or some other place where store managers actually know customers by name and by sight.

Submitted by Chuck Jeffries of Greensboro, North Carolina.

A complimentary book goes to Stephen H. Cook of North Providence, Rhode Island, and to Joseph S. Blake, Jr., of Ottawa, Kansas. Thanks also to the scores of readers who made the same points as our grocery experts but whom we did not have space to quote.

FRUSTABLE 10: ***Why do so many policemen wear mustaches?***

For some reason, civilians were reluctant to speculate about this issue, but we heard from quite a few policemen and friends and relatives of cops. We were surprised that so many responses echoed that of Scott Miles of Santa Monica, California:

> My best friend, Tim, is a police officer. Like many of his fellow officers, Tim has what some would call a "baby face." His features are very soft and his looks lead people to assume, initially, that he is a good-natured, easygoing guy.
>
> In Tim's case, this impression is true, but it is not a very good image for a police officer who needs to command authority. Tim grew a mustache to harden his look, to make him look a little older and tougher.

Of course, some prefer to look older for personal reasons. A few officers repeated the sentiment of Eric Crane of Texarkana, Texas: "My wife won't let me shave mine off because I would look like I'm sixteen years old."

Many readers and police officers mentioned that in our culture, mustaches mean macho. We are left with a chicken/egg problem: Do cops grow mustaches because they are already macho? Or do they appear more macho because they wear a mustache? Clifford Smith, an officer in Anne Arundel County, Maryland, notes that mustaches date back to the very first "cops" in Robert Peel's London service, and that "staches" have always been associated with military and paramilitary men.

Although a young officer might want to project machismo, we're guessing that other factors are more important. Most of the military and police officers we've met, despite the visual image they project, want to express their individuality as persons. Confronted with strict controls over what they can wear and how they can look, a mustache becomes a way to stand out from the crowd. Many cops concurred, among them Ohio police officer Mike Foley:

> We have to have short hair and no beard. The *only* allowance for any "style" is a mustache. Most cops (after a few years) want to look like something other than a boy scout, so they grow one. Plus, it keeps the bosses mad when you don't trim it.

Foley's combination of humor and mild rebellion was typical of almost all of the letters we received from police officers. Noncop Bob Kowalski of Detroit, Michigan, aptly summarizes the feelings of many police officers who wrote us:

> Perhaps policemen would not be so keen to have mustaches if police departments weren't so keen on banning them! The old forbidden fruit . . .

And we were moved by this note from the policeman's bulletin board on Prodigy Service. Leo Martin raises a point we had not considered—some cops might tire of looking like cops:

> I am a cop out in northern California, and I've had hair growing from my lip up to my nose for most of my life. It started in the Marine Corps, where the only facial hair you could have was a "stache."
>
> Most police departments have the same policy. I think growing a mustache makes us feel and look more like regular people and less like cops.

We're faced with a paradox. Cops grow mustaches to express their individuality, yet so many cops wear them that it defeats the purpose. This irony was not lost on our favorite self-examination, from admitted baby face Steven J. Schmidt of Covington, Kentucky:

> As a mustache wearer for eighteen of my nineteen years as a policeman, I have some insight into the question. The obvious answer to this Frustable is, "Because they can." Although police departments aren't as paramilitary as they used to be, most policemen cannot wear beards or grow their hair long. A mustache, therefore, becomes a symbol of their individuality.
>
> I know that makes about as much sense as all the people in the 1960s who grew their hair long to demonstrate their "individuality," but there it is. You might notice that a lot of retired policemen and discharged servicemen grow their hair long or grow a beard right after leaving "uniformed" service "just because they can." Once the novelty wears off, they revert to what's comfortable for them.

Schmidt reports that once he turned forty, "I didn't need the mustache to look older anymore. In fact, I cut it off to look younger."

As we mentioned, the cops we heard from demonstrated quite a sense of humor. Here are some of the, er, more unusual solutions to this Frustable:

> I have a "stache" just because it gives me something to do while I drive around for eight hours. You can twist it and pull it for eight hours and you smoke fewer cigarettes. Sounds silly but it works for me. (Bill Pador, Jr., New Jersey)

> I like to make my partner feel like I'm one of *him.* You may not realize that I'm referring to my partner, Max, who is a horse, and also has hair around his lips. (Christopher Landry, New Orleans, Louisiana)

> Now that you mention it, my mustache probably came from off the top of my head. (Greg Wilson, Delaware)

> I guess it was so my kids could pull it when it was time to get up for work. (Alan Levine, Philadelphia)

And, last but not least,

> You can lick the doughnut powder off a mustache for hours after the last coffee break—and it still tastes good. I usually cut mine off in summer because if I don't, everyone knows I've been eating ice cream. Also, a "stache" may act as the last air filter in the air we sometimes have to breathe—I go home and hose it out. Lastly, if a catfish can have one, so can I. (William Howe, IV, Prodigy Service)

Submitted by Steve Propes of Long Beach, California. Thanks also to Laura Arvidson of Westville, Indiana, and Brad Huddleston of Bakersfield, California.

A complimentary book goes to Steven J. Schmidt of Covington, Kentucky. Thanks to all the officers from the Police Topic on the Prodigy Service.

The Frustables That Will Not Die

Imponderables readers don't give up. Even though we have partially flushed out the frustration posed by the Frustables in other books, readers want to leave no stone unturned. Your most recent contributions to the demolition of past Frustables are presented here.

Please remember we do not have the space to review all the theories we've already advanced; this section is meant as a supplement, not a substitute, for our discussions in previous books.

Frustables First Posed in *Why Do Clocks Run Clockwise?* and First Discussed in *When Do Fish Sleep?*

FRUSTABLE 1: ***Why do you so often see one shoe lying on the side of the road?***

In the past, we've heard from mostly urban SSS (single-shoe syndrome) theorists. But there has been a recent rash of sightings in more bucolic settings, like this one from Donald Mueggenborg of Lemont, Illinois:

> We marathon canoe racers often get our feet wet. One year, it became fashionable to tie our wet shoes to the rear thwart when transporting canoes. The shoes not only dried out but became sort of a sign of fellowship. Often, one shoe would become untied and fall off.
>
> Is there any good way to use one tennis shoe?

Sure, if you are Alyssa Constantine, of New York City. Alyssa lives in a woodsy area of Queens, where hikes and field trips are often conducted by nearby schools, with

> . . . students carry bulging backpacks, heavy sleeping bags, and a tote full of life's necessities. Every forty minutes or so, the teacher allows students to take off their shoes and pamper their aching feet.

> Every day, I come out and collect forgotten, mateless shoes, lost in the sea of jackets, totes, sleeping bags, etc. I now have at least one hundred different shoes in my basement.

One hundred shoes in the basement? Whatever happened to collecting dolls? Or stamps? We still don't think Alyssa's stash is sufficient to explain the widespread sightings of single shoes on the road. Perhaps the answer could lie in perverse camp counselors? We heard this startling testimony from Mary Beth MacIsaac, from beautiful Cape Breton Island, Nova Scotia:

> When I was nine (that was only two years ago), I went to Brownie camp. They had a contest among the cabin mothers to see who could kick off their shoes the farthest. Many of them lost the shoe that they had kicked off. If every Brownie and scout camp did that every year, you can just imagine how many shoes that is.

The mind shudders.

But American citizen Regina Earl thinks that the SSS conspiracy is broader than infiltrating scout camps. Regina has recently come back from a stint in Japan, where she saw an ominous shoe commercial, featuring Charlie Sheen, of all people:

> Charlie Sheen is shown lying in the center of a long, deserted road holding one shoe, while a girl drives away in a red convertible holding the other. I know this doesn't provide any answers, but it shows Frustables jumping international borders.

That's okay, Regina. We must confront the scope of a problem before we can solve it. In this spirit, Wayland Kwock of Aiea, Hawaii, was brave enough to share a harrowing article found in the November 1992 *Scientific American*, announcing that on May 27, 1990, a freighter dumped five shipping containers containing eighty thousand Nike shoes into the waters of the Pacific Northwest.

Several scientists decided that the accident afforded them a unique opportunity to study ocean drift. At last report, some of the shoes are heading for Japan (sorry, too late to retrieve them, Regina). Amazingly enough, none of the scientists tracked how many Nikes ended up, singly, on the side of a road.

FRUSTABLE 2: ***Why are buttons on men's shirts and jackets arranged differently from those on women's shirts?***

We are still receiving loads of mail on this button Frustable. Most people debate the relative merits of what we've already published. But we did receive one new theory this year from Terri Longwell of Richland, Washington:

> Unless I was lied to at a young age, I learned the answer to this question as a sophomore in my high school American history class. The answer dates back to the garment sweat shops in the days before unions. The "piece workers" were paid more for sewing women's garments than for men's. . . . To separate the men's blouses from the women's blouses, which in those days were similar, the buttons were sewn on opposite sides. . . . The employees who were favored by management got the better, more high-paying pieces, which were the women's blouses.

Terri, we think you were probably "lied to." In what era were women's blouses indistinguishable from men's shirts? Weren't the collars totally different?

FRUSTABLE 4: ***Why do the English drive on the left and most other countries on the right?***

Readers can't stop debating this Frustable, either. We have nothing new to say, other than that we despair of ever getting a definitive answer. If you want to become more informed, if perhaps more confused, you might want to read a whole book on the subject: *The Rule of the Road: An International Guide to History and Practice,* written by Peter Kincaid (Greenwood Press, 1986). Thanks to reader John P. Hersh of Concord, Massachusetts, for recommending it.

Frustables First Posed in *When Do Fish Sleep?* and First Discussed in *Why Do Dogs Have Wet Noses?*

FRUSTABLE 6: *Why do so many people save National Geographics and then never look at them again?*

Mike Teige of Seattle, Washington, tipped us off to the latest literary take on this (literally) weighty subject. In his 1992 book *Clutter Control,* cleaning expert Jeff Campbell devotes an entire chapter to the topic of controlling rampant National Geographic proliferation. Campbell's chapter heading, "They seem to be everywhere," echoes the sentiment that inspired this Frustable:

> As of September 1991 there were approximately 3.784 *billion* copies of *National Geographic* in print, and just about every one of them is still in somebody's garage.

Unfortunately, Campbell provides many more tips about how to get rid of the magazines than theories about why we didn't throw them away in the first place. Still, if Campbell can inspire enough people to get off their duffs and recycle *National Geographic,* we could at least eradicate the intellectual clutter created by this Frustable.

FRUSTABLE 7: *Why do people, especially kids, tend to stick their tongues out when concentrating?*

Reader James D. Kilchenman of Toledo, Ohio, was kind enough to pass on the information that in his recent book, *Babywatching,* anthropologist Desmond Morris weighed in on our debate about this Frustable. Morris notes that children, when rejecting food (either solid, a bottle, or mother's breast), stick out their tongues to push away the food source.

Morris believes that sticking out the tongue is a universal rejection signal, even when used unconsciously. When an adult is concentrating on a difficult task and sticks out the tongue, the tongue is behaving just as it did when, in infancy, it rejected an insistent parent offering food. The message now, as then, is the same, namely: "please leave me in peace."

While we think that the physiological theory presented in *Why Do Dogs Have Wet Noses?* is compelling, Morris's explanation also makes some sense. But we were also taken aback by a letter from reader Mark McGrew of Bucyrus, Ohio, who offers an alternative:

> Human kids are not the only kids to stick out their tongues when they are trying to concentrate. In the Jane Goodall documentary "People of the Forest," you can see a young chimpanzee trying to twirl around, which he can do well only after he begins sticking out his tongue. Perhaps we simply inherited the practice from them.

Frustables First Posed in *Why Do Dogs Have Wet Noses?* and First Discussed in *Do Penguins Have Knees?*

FRUSTABLE 1: *Does anyone really like fruitcake?*

We'd like to think that we have been duly modest in boasting about attempts to edify our readers. After all, we're not dealing with metaphysics here. Still, every once in a while, we receive a letter that lets us know we have made a profound difference in someone's life. We received a moving letter from Howie Saaristo of Norfolk, Massachusetts, indicating that our modest efforts have changed his life unalterably for the better:

> For years I ordered and gave away more than a dozen fruitcakes at Christmas. Then I read that you asserted that most people do not like fruitcake. I was completely astonished! I like fruitcake so much that such a thought had never occurred to me.
>
> After some thought, I decided that I had better ascertain the truth. I called each one of my former recipients and told them what I had read and pleaded with each to tell me the truth.
>
> Out of the lot, it turned out that I was the only one who really liked the stuff. The others just suffered in silence. I put a stop to giving them away. What a shame they don't know what they are missing!

After such a great start, Howie, we're worried about that last sentence. We suggest a local twelve-step program.

According to Julia Ecklar of Monroeville, Pennsylvania, the answer to whether anyone really likes fruitcake is: "Yes, but only when the fruitcake aficionado has a physical problem." Julia can think of only two possible reasons for her father's strange predilection: One, he loves rum, particularly hot buttered rum. Many fruitcakes contain rum, which, as Ecklar so felicitously puts it, supplies the cake with "its strong, odious flavor." But we're more concerned with her second explanation:

> My father has no sense of smell. As far as I know, he never did, and it has certainly affected his ability to taste. . . . This lack of taste discrimination might also have contributed to his liking of fruitcake.

We're a little upset at the Ecklar extended family for taking advantage of his handicap, for Ecklar reports:

> When I was a kid, all our relatives knew that giving fruitcake to my father would just make his holiday; all our friends knew that they could dispose of unwanted fruitcakes by giving them to us.

Christopher K. Degnan of Whitefield, New Hampshire, theorizes that the reason fruitcake is so unpalatable is that most cakes withhold *the* crucial ingredient: pork. Christopher shared with us a recipe contained in a book called *Vermont Cooking*. The recipe for two loaves includes the usual nuts, molasses, raisins, fruits, eggs, and sugar, as well as "one cup of chopped pork (all fat)." The mind reels.

FRUSTABLE 4: *Where, exactly, did the expression "blue plate special" come from?*

Several correspondents were aghast at the readers quoted in *When Did Wild Poodles Roam the Earth?* who doubted that blue plates were ever actually used in restaurants. Our favorite was Dave Rutherford of Holcomb, New York, who sent us a color snapshot of a classic diner, the Miss Albany Diner in Albany, New York. Right next to the name on the sign above the door was another sign: "Blue Plate Specials." Dave reports that specials advertised on the sign "were and continue to be served on blue plates." Judy Stuart of DeLand, Florida, enclosed an article about the blue willow plates we wrote about in our initial discussion of this topic in *Do Penguins Have Knees?* But Judy has more concrete evidence that blue plates were used in restaurants: Her husband, Dick, worked in diners in the late 1930s and early 1940s and served many a meal on blue plates.

Indeed, we heard from several readers who collect blue willow china. We particularly enjoyed a long discussion by Pat Kaniarz of Harbor Springs, Michigan. Kaniarz confirms that the earliest blue and white china was imported, appropriately enough, from China, and was all hand painted.

With the advent of the Industrial Revolution, English potters figured out how to transfer print designs stolen from uncopyrighted Chinese design. Eventually, the English mass-produced willow patterns in a myriad of colors, but blue and white was always the most popular.

Most blue plate specials were served on divided dishes called "grill plates." According to Kaniarz:

> Those grill plates, once so inexpensive that restaurants let low-paid dishwashers handle them, now are offered in antique shops at twenty to thirty dollars each. Lots of restaurants used blue willow, although not all confined their use to the grill plates.
>
> The restaurant ware is very collectible. Some of us who are hooked on blue willow (we have a newsletter called "The Willow Word" that is subtitled "The Newspaper for People Addicted to Willow-Pattern China") collect only the restaurant ware. Because it was made for heavy use, a lot of it has survived in pretty good condition.

In the three books since *Why Do Dogs Have Wet Noses?*, we have pieced together the origin of the blue plate special, but our initial target still eludes us. We still haven't found *the* restaurant that initiated or inspired the expression.

FRUSTABLE 7: *Why and where did the notion develop that fat people are jolly?*

In *Do Penguins Have Knees?*, we mentioned Shakespeare's Falstaff, the archetypical fat-jolly person. But reader Judith Goldish of Lakewood, California, reminded us of another of the Bard's pronouncements, from Julius Caesar:

> Let me have men about me that are fat;
> Sleek-headed men and such as sleep o' nights.
> Yond Cassius has a lean and hungry look;
> He thinks too much: such men are dangerous.

FRUSTABLE 8: *Why do pigs have curly tails?*

Buhnne Tramutola of Annandale, New Jersey, an ex–pig raiser, confirmed what we have written before—that a curly tail is an indicator of healthiness and happiness, just like a dog's wagging tail. If the pig's tail was not curly, we would check into its health or living conditions.

Old news. But Tramutola has something new to contribute. Pigs' tails were once used to grease pancake griddles. According to an unidentified book he sent, a pig's tail "would last for weeks if kept cold. But mostly the extra fat was used in soap making." That would be enough to curl our tails.

FRUSTABLE 9: *Why does the heart depicted in illustrations look totally different from a real heart?*

We thought we exhausted the possibilities in this question. But we were wrong. In *Do Penguins Have Knees?*, we mentioned Desmond Morris's theory that our Valentine's heart is an idealized version of the female buttocks; reader Kierstyn Piotrowski of Parsippany, New Jersey, with the help of Kassie Schwan, presents a similar, ingenious theory:

> If you put the profile of a man and woman in a "kissing position" (excluding the inevitable turning of heads to avoid nose bumping), it looks roughly like this.

Jerry Tucker of Burton, Michigan, claims that the secret to this Frustable has been unlocked for centuries by fellow Native Americans:

> Take a walk in the woods. Any bush or shrub that has a leaf shaped like the Valentine heart has medicinal qualities especially beneficial to the human heart.
>
> We have for centuries identified medically beneficial shrubs and bushes by the shapes of their leaves. This particular shape was adopted by white people to represent their concepts of love, romance, etc.

We're not sure, though, how this theory accounts for the spread of the "leaf-shaped" heart to non–Native Americans. Tucker presented us with some leads to confirm his theory, and we'll report back if we find out more.

FRUSTABLE 10: ***Where do all the missing pens go?***

Two readers have taken us to task for our secular-humanist explanations for the disappearance of pens. The answer, they insist, lies in felonious felines. To wit: Here is the sworn testimony of Rainham D.M.H. Rowe of Jacksonville, Florida:

> One morning I was faced with the task of finding my wedding rings, after I had left them on the kitchen counter the night before. I happen to have three cats, two of which are notorious for climbing on the counter, where they know they aren't supposed to be, to find things to play with.
>
> One cat in particular loves the little rings that come off milk jugs. I figured this cat must have seen my rings and thought they were milk jug rings and knocked them off the counter to play with.
>
> I began my search by shining a flashlight into every crevice in the kitchen, to no avail. I then pulled out the appliances. Under the range I found a handful of magnetic ABCs, about ten milk jug rings, and lo and behold, *five pens.* There was a similar sight under my refrigerator.
>
> I eventually found my rings under the computer desk, and found *another handful of pens,* piles of paper that had fallen out of the back drawer, and a toy car. So if *Imponderables* fans have cats, perhaps their pens are being used as nocturnal entertainment.

Rainham, you won't convince Janet Sappington of Hope Mills, North Carolina, otherwise. In her cats' "hidey holes," she has found numerous pens, as well as pen caps, coins, lighters, socks (oh, that's where the missing socks are!), and once, a whole shirt.

Frustables First Posed in *Do Penguins Have Knees?* and First Discussed in *When Did Wild Poodles Roam the Earth?*

FRUSTABLE 1: ***Why do doctors have bad penmanship?***

We thought we exhausted this topic in *Poodles*, but two readers raised points that we never considered. David A. Crowder of Miami, Florida, stresses that penmanship is not stressed or valued highly in schools, particularly for boys:

> As a child, I got excellent grades in all subjects but penmanship. My parents, who would have hit the ceiling had I gotten less than a B in any other subject, would shrug at a D in penmanship—after all, their son was going to be a doctor!
>
> With no pressure to perform in this subject, and no apparent benefit otherwise, I would never put much effort into it.

We doubt if penmanship is highly prized in medical school, either. But Crowder feels that there may be neurological reasons why doctors might tend to have poor penmanship:

> As Betty Edwards points out in her seminal book *Drawing on the Right Side of the Brain*, "you can regard your handwriting as a form of expressive drawing." That is, there is an artistic form to handwriting, an indication of, among other things, artistic ability and perception. Since good artwork is predominately right-brain activity, it is not surprising that any sort of scientist or technician, whose life, work, experience, and study involve mainly left-brain activity, would be deficient in a right-brain function (after all, who can be good at everything?).

Crowder indicates that he never became a doctor but that his "lousy handwriting" led directly to his career in computers. We can testify though, after seeing David's signature, that he would have made a *fine* physician.

In *When Did Wild Poodles Roam the Earth?*, we mentioned that on occasion physicians might deliberately attempt to obscure their handwriting for relatively benign reasons. But Rose Marie Centofanti of Chicago, Illinois, offers a far darker scenario:

I am a member of a profession called "health information manager." Part of our professional responsibilities entails the legalities of medical records.

The documentation in a medical search is primarily the responsibility of physicians. If a case is brought before a jury in a court of law, the medical record may be subpoenaed as evidence. The physicians will also be subpoenaed and have to read aloud their documentation as testimony in court.

Because their penmanship is illegible in most cases, they can state they've written just about anything.

FRUSTABLE 3: ***Why don't people wear hats as much as they used to?***

Several readers wanted to add another motivator for uncovered pates, at least for females: the Catholic church. Typical were the remarks of Vega Soghomomian of Maple Grove, Minnesota, who wrote:

> In the Catholic church, females were *required* to wear a head cover. (We would not want to offend God!) If a woman forgot her hat, she wore a hanky or even a piece of tissue paper on her head. With the women's movement, the Catholic church removed the rule and women went out and bared their heads to the world.

FRUSTABLE 4: ***How and why were the letters B-I-N-G-O selected for the game of the same name?***

Not too much to report here, other than a fascinating theory by Rick Biddle, president and general manager of WOWL-TV, an NBC affiliate in Florence, At The Shoals, Alabama. Biddle was once responsible for producing, directing, and starring in a television bingo show, and heard from a bingo supplier that the expression in question is an acronym:

> Think of what it would be like if you filled all the numbers on your bingo card. At the same time, three or four other people filled in the numbers on their cards and you all jumped up simultaneously yelling, "I've placed all the little balls in the holes with corresponding numbers and have won the game!"

> Rather than going through this rather lengthy dissertation, the word bingo is derived from, "*B*alls *I*n *N*umbers *G*ame *O*ver."

We know for a fact that the original bingo markers were not balls, but beans and seeds, which makes this theory less than likely to be true.

FRUSTABLE 8: ***How did they measure hail before golf balls were invented?***

We received, pardon the expression, a flood of letters about this Frustable in the past year. Most were variations on the analogies to edibles (e.g., peas, eggs, walnuts) we discussed in *Poodles.* But four different readers directed us to what might be the first written reference to the size of hail in the Western canon, Revelation 16 verse 21: "And great hail from heaven fell upon men, each stone about the weight of a talent."

According to reader L. Ray Black of Arcadia, Florida, the talent was the largest of the Hebrew units of weight. Two Davids from Keizer, Oregon, Messrs. Engle and Volkov, indicated that the talent fluctuated over time from fifty to one hundred pounds. Reader Dale Gilbert of Chillicothe, Ohio, adds that the Bible indicates that hailstones provoked men to blaspheme God, "for the plague thereof was exceeding great."

Now that *Imponderables* books are being published overseas, we are starting to hear reports from far-flung ponderers. Emmanuelle Pingault reports that in France, hailstones in weather reports are invariably, if unsurprisingly, compared to foodstuffs.

> A small hailstone may be compared to a nut (in French, *noisette*), while a larger one will be "as large as a walnut" (*gros comme une noix*). But the more frequent set of comparisons, regularly heard in weather reports, is as follows: as large as a pigeon's egg; as large as a hen's egg; and larger than a hen's egg.

Emmanuelle wonders how many of us have actually ever seen a pigeon egg. We now know why Imponderables are universal—the seeming nonexistence of baby pigeons was one of the original inspirations for our first book, *Imponderables.*

But all the other efforts of readers pale before the research undertaken by reader Chip Howe of Washington, D.C., who conducted a

Nexis database search of newspaper weather reports and proved conclusively that we sorely overestimated the ubiquity of "golf ball" analogies to the size of hail. We're proud to announce that literary imagination is not dead. Here are the categories and some of the quotes that Chip found in newspapers over the last six years:

- *Sports:* softball-sized; tennis ball-sized; baseball-sized; golf ball-sized, of course; marble-sized; and ping pong ball-sized.
- *Food:* grapefruit-sized; orange-sized; lime-sized; cherry-sized; egg-sized; walnut-sized; "hail the size of Spanish olives"; bean-sized; pea-sized; butterbean-sized; ice cube–sized; and our personal favorite, dry roasted peanut-sized hail, from Georgia, the peanut state.
- *Money:* Chip could almost start a coin collection. He found references to every current denomination of American coinage except the silver-dollar.
- *Body Parts:* "Hail the size of babies' toes" and this scary report from Canada: ". . . after a torrential thunderstorm had pelted Edmonton with fist-sized hail stones."
- *Two-for-one:* pea- to marble-sized hail and quarter to tennis ball-sized hail.
- *Nature's Own:* acorn-sized hail; pellet-sized hail; pebble-sized hail; mothball-sized hail; and a contribution from Bulgaria (snowball-sized hail), a country that must have more uniformity of snowball size than we do in North America.

FRUSTABLE 10: ***Why does meat loaf taste the same in all institutions?***

When we first posed this Frustable in *Do Penguins Have Knees?*, we mused: "Does the government circulate a special Marquis de Sade Cookbook?" One reader, Carl Bittenbender of Staunton, Virginia, answers "yes":

> Many institutions of all kinds, college, military, hospitals, etc., use the armed forces menu cards, which give recipes for cooking hundreds of meals at a time. The cards give formulas for figuring the amounts of ingredients needed for large recipes. This accounts for the bland, tasteless quality of many of the recipes, as they are designed to be eaten by people who do not have a choice.

Reader George E. Jackson, Jr., of Mantua, Ohio, notes that the federal government supplies many institutions with surplus food and that they are

> extremely picky about their suppliers meeting stringent specifications. . . . As an example of what I mean: the military recipe for fruitcake is eight pages in length. Perhaps if you contacted the Government Printing Office, you might be able to get their recipes for meat loaf and fruitcake.

Sorry. We'd rather write the IRS, asking them to please audit us.

Of course, the government has its civilian culinary counterparts in the large institutional catering companies, such as Marriott and ARA. As Mike Tricarico, Jr., of Dubuque, Iowa, puts it:

> The meat loaf you had last month for lunch at a hospital in New York was very possibly made by the same company, following the same recipe, as the meat loaf that you ate yesterday at an IBM cafeteria in southern California. Marriott also serves food on airlines, so it is even possible that you had the very same meat loaf on board flight 123 from New York to southern California. Hopefully, this entree was not followed by fruitcake!

Is it our fate for fruitcake to follow us everywhere? We're talking meat loaf, now.

Only one reader was willing to plumb the ineffable essence of meat loaf. And that savior is Wayland Kwock of Aiea, Hawaii:

> It boggles the mind that nobody would be brave enough to expose the meat loaf conspiracy. Closer inspection would probably show that meat loaf is served on Friday, the end of the week. The day to get rid of all the "extra" food, the dregs, the leftovers. All of this goodness is unceremoniously included in the meatloaf.
>
> So why does it taste the same? Mathematics, specifically probability, provides the answer. If an infinite number of monkeys . . . No, that's not quite right. If an infinite number of institutions served an infinite variety of food, the amounts and types of leftovers would tend to form a Gaussian distribution. This means that there may be meat loafs out there that taste better (not better—different), but they are outside one, if not two, standard deviations. All other meat loafs contain an average amount of a generic sampling of foods and thus, on average, taste the same.

We don't understand a word that Wayland says here, but we smell greatness. Or is that fruitcake we smell?

Tuesday
Dear Dave ~
I believe I
have the
nswer to
24815
US MAIL
LETTERS

Imponderables *readers have continued to flood our post office box with thousands of letters in the past year. We appreciate all your new Imponderables and solutions to Frustables. And we wouldn't be human if your words of praise didn't put a spring in our step. But this section is reserved for those of you who have a bone to pick with us: Some of you want to add to what we've discussed; others want to disagree with what we thought were words of wisdom.*

Please remember we can publish only a fraction of the terrific letters we receive. Many of you have submitted corrections or suggestions that we will be researching; we will check out your concerns even if we don't publish your letter. Because of the mechanics of publishing, it can sometimes take years to validate objections and change the text on subsequent printings, but we do so regularly. The letters contained here are chosen for their entertainment value and the merit of their argument. Let the bashing begin!

Is it Clintonomics? The coming millennium? We don't know why, but Imponderables readers were particularly testy this year. Sometimes for good reason. Several readers, such as Jon A. Kapecki of Rochester, New York, took us to task for our discussion of peanut M&Ms:

> Nuts may indeed be "the source of one of the most common food allergies," as you assert on page 56 of *Do Penguins Have Knees?* However, peanuts—the subject of discussion—are not nuts, but legumes, specifically members of the pea family.
>
> This is no pedantic distinction. People who are allergic to nuts are usually not allergic to peanuts and vice versa, and a failure to observe such distinctions can have fatal consequences.
>
> That said, I enjoyed the book.

Gee, Jon, that's a little like saying other than being mass murderers, we have a pleasant personality. But you are right. Peanuts are not technically nuts, and we should have been more careful in our terminology. Violent allergic reactions both to peanuts and other nuts are common, but someone who reacts to pecans or walnuts may suffer no adverse effect from consuming peanuts.

Speaking of getting sick, two more faithful readers and correspondents, Rabbi Joseph Braver of Baltimore, Maryland, and Fred Lanting of Union Grove, Alabama, wrote to complain that our dis-

cussion of the snake emblem found on ambulances in When Did Wild Poodles Roam the Earth? *was woefully incomplete. For the associations of the snake and the pole have Jewish and Christian as well as Roman and Greek significance. In Numbers 21, the wandering Israelites were afflicted by snakes sent by God to punish them for speaking against Him. Moses interceded on behalf of his suffering followers:*

> Then the Lord said to Moses, "Make a seraph figure and mount it on a standard. And if anyone who is bitten looks at it, he shall recover." Moses made a copper serpent and mounted it on a standard; and when anyone was bitten by a serpent, he would look at the copper serpent and recover.

Fred Lanting points out that the snake and staff symbolism continues in the New Testament and asserts that "the Greek myths were corruptions of the stories of Israel's experiences with this Old Testament healing.

While we're on the subject of vehicles, readers are still trying to figure out where the old oil lurks in automobiles after oil changes. In Do Penguins Have Knees? *we could account for much but not all of the disappearing oil. We heard from Dan Kiser of Elmira, New York, a student studying automotive technology. If you combine Dan's account with our previous discussion, we think this Imponderable is finally nailed:*

> Assuming that the engine is warm and that it does indeed have five quarts of oil, here is where the oil "lurks." Most all engines are made out of cast iron and manufactured by a process where cast iron is poured into a sand mold. This process creates a rough texture on the surface of the engine block. Oil will cling to this surface because of the rough texture of the block. Oil will settle mainly in the lifter galley (located directly under the intake manifold). It will also accumulate on the top surfaces of the cylinder heads.
>
> The crankshaft and connecting rods in your engine ride on a thin film of oil between two bearing halves. If oil was not present here all of the time, your engine would self-destruct due to lack of lubrication. There is about .003 clearance between these bearing halves and the crank or rods. This oil will not drain out during an oil change.

Oil will also stay inside the oil pump and oil pump pickup during an oil change. The lifters in an engine operate the pushrods, which in turn open the valves. These lifters (one for each valve, sixteen in a V-8 engine) are of the hydraulic type: They are filled with oil during their lifetime. This oil will not drain during an oil change, either.

If you were to drain the oil out of your engine and then put the drain plug back in, you would have to wait several weeks before any of the oil in the aforementioned spots drained into the pan. . . .

I have had the opportunity to rebuild several engines. In each case, the oil was drained and the engine was allowed to sit for several weeks. By the time I started to disassemble the engine, residual oil had drained into the pan, resulting in about three-quarters of a quart of oil on my garage floor when I pulled the oil pan off.

Oil wasn't the only liquid on your minds over the past year. Many of you are concerned about water, in particular, bodies of water. In Do Penguins Do Knees? *we discussed the difficulties in differentiating a lake from a pond. Several readers insisted there was a distinction. Typical was this letter from Bill O'Donnell of Eminence, Missouri:*

As an ecosystem, a pond is defined as a body of water of such a depth that light can penetrate all the way to the bottom, allowing rooted submergent vegetation to grow across the entire bottom. Lakes are deeper, so that rooted plants cannot draw at the deepest parts. They also have differences in temperature, called thermoclines, which ponds usually lack. Of course, many true ponds are called lakes and vice versa, but as you said, people can get away with calling most things anything they want.

Exactly, Bill. We're talking apples and oranges. We were discussing geographical definitions, and you are speaking of biological ones.

More liquids? In When Did Wild Poodles Roam the Earth? *we discussed what we are smelling when it "smells like rain." Ron Smith of Winnipeg, Manitoba, wants to supplement our explanation:*

Just before a storm, the barometric pressure decreases. Rising air reduces surface pressure and produces condensation, quite often resulting in cloud formation and frequently precipitation. The reduced surface pressure causes slight gas release from the soil resulting in a fresh or "earthy" smell.

A fellow Canadian, Gilles Fournier of Calgary, Alberta, wanted us to know that in local folklore, the "H" in the "C" of the Montreal Canadiens stands not for "hockey" but for "habitants":

In the English media, the Montreal Canadiens are often affectionately called "the Habs." Most people believe that Habs is short for "habitants," the French word often used to mean "farmer" in Quebec. What do farmers have to do with hockey? Many of the Canadiens wunderkinder came and still come from the rural areas of the Belle Province . . . so from Habitants, to Habs, to . . . the H in the C!

In Why Do Dogs Have Wet Noses?, *we discussed why there is no channel 1 on televisions. Gilles wanted to add:*

the FCC (and its Canadian counterpart, the CRTC) gave it back to radio buffs because channel 1 was a poor TV performer, riddled with ghost images. That frequency was just too prone to interference from other radio frequencies.

Most of your complaints this year have been about our discussions of technology. We can always count on a few correspondents to offer constructive criticisms about our explanations of gadgets and widgets. One of our more irrepressible contributors is William Sommerwerck of Bellevue, Washington:

When Did Wild Poodles Roam the Earth? states that 9-volt "transistor" batteries are rectangular because they take the shape of six stacked cylindrical cells.

This is absolutely, utterly, completely, and *totally* wrong, wrong, Wrong, *Wrong*, WRONG!!!

The cells in a 9-volt battery are rectangular. They look like little sardine cans, but (as a friend said) without the key. If your so-called "expert" had ever bothered to open one, he or she would have seen this.

But how do you really *feel about our discussion, Bill? You motivated us to call back several battery companies, and all we can tell you is that if the cells of 9-volt batteries are rectangular, the technical staffs at Eveready, Duracell, and Panasonic don't know about it. Eveready's 9-volt alkaline battery, for example, contains six quad-A cells, which are now being marketed separately as E-96 batteries and are used primarily in penlight flashlights and laser pointers. Perhaps, William, you are thinking about less popular carbon-zinc batteries, which often contain rectangular or "cake" cells stacked atop one another inside the case.*

Believe us, William Sommerwerck isn't our only correspondent with a bee in his bonnet. By far the angriest and most vociferous mail we received this year came from the eight readers who violently objected to a letter we published in Poodles *about why tape counters on audio and video tape players don't seem to measure anything. We quoted an electronics engineer who claimed that tapes didn't run at a constant speed. Thank you, Stan Sieger, Michael Javernick, Dallas Brozik, Nils J. Dahl, Jr., Charles Kluepfel, Jim Tanenbaum, and Bruce Hyman for setting us straight; but we'll quote the letter from John B. Dinius, of West Hartford, Connecticut, because his explanation is simple enough for even us technoramuses:*

> All audio tape recorders (with the possible exception of some really cheap models that would be considered toys) move the tape past the heads at a constant speed, by using a capstan and pinch roller. The function of the takeup reel is not to control the speed of the tape but merely to collect the tape after it passes the capstan/pinch roller device. This constant tape speed is evidenced by the fact that technical specifications for tape recorders always express the tape speed in terms of inches per second (e.g., cassette tapes play at 1-7/8 inches per second).
>
> Your correspondent suggests that the tape passes the read/write heads of a VCR faster towards the end of a movie because the effective diameter of the takeup reel has been increased by the tape that has been collected. In fact, you can observe (by noting the number on the counter every fifteen minutes while the tape is running) that the tape counter runs more and more slowly as the movie progresses. This indicates that the takeup reel has to turn more and more slowly in order to collect the tape, which is moving past the heads at a constant speed.
>
> As far as the original question is concerned, the reason [why tape counter numbers seem arbitrary] is that they measure revolutions of the takeup reel, which don't bear a constant relationship to the things that people really care about (i.e., how many minutes into the tape they are, and how much time is left on the tape). Note that if the reel actually ran at a constant speed, as your correspondent suggested, then

the number of revolutions would be proportional to the elapsed time of the tape, and people could use the counter numbers fairly well, by realizing that a certain number of revolutions represented one minute of tape.

As Dinius mentions later in his letter, fortunately for us, most VCRs now use time counters, which measure information much more important to the average consumer. Our next angriest group of correspondents challenged the comments of a source in When Did Poodles Roam the Earth? *who discussed why trees on a slope don't grow perpendicular to the ground. Our first correspondent is Stanley Sieger of Pasadena, California:*

> Hardly one of his sentences is without error *or worse.* He attributes "motivation" to trees, claims that light provides trees with food (rather than just the energy to "digest" the food they absorb), claims that in a forest the source of light is "up," etc.
>
> But worse, oh so much worse, is his confusing geomagnetism with gravitation and claiming that there are places of "abnormal" gravity on this planet. Wrong!

The original reference was to the Oregon Vortex in Gold Mill, Oregon, where our source said that "it is reported" that trees grow in a contorted fashion because of abnormal gravitational forces. Scot Morris joins Sieger in (justifiably) abusing us for allowing these statements to go uncriticized. Morris, a regular contributor to Games *magazine, personally conducted an investigation of the Oregon Vortex, published in the December 1987 edition of* Omni, *and proved that this place where balls that appeared to roll uphill was clearly an illusion.*

Another reader, David A. Crowder of Miami, Florida, has a bone to pick with another one of our sources, who in Do Penguins Have Knees? *claimed that surge protectors can protect your VCR from damage during lightning storms:*

> A surge suppressor will only protect against minor power surges and spikes such as commonly occur in any electrical line. . . . Lightning, though, is far more powerful than anything any surge suppressor or line stabilizer is capable of handling. A lightning strike will simply blow the surge suppressor as it fries your VCR or computer.

While we're on the subject of things technological, Howard L. Helman of Manhattan Beach, California, rightfully comments that our discussion of "Where do computer files and programs go when they are erased?" was correct for MS-DOS machines but not necessarily for other operating systems:

> As far as I know, MS-DOS is the only system that uses the signa character to mark the deletion. Other systems usually have a flag in the directory to mark the entry free.

While we're speaking of computers, we heard from Harold Gaines of St. Louis, Missouri:

> I must comment on your response to the question, "Since computer paper is longer than it is wide, why are computer monitors wider than they are long?" Your answer is fine, except that an important fact was overlooked. The reason virtually all computer displays are eighty characters wide (especially terminals on multiuse systems) is because the first popular highspeed input devices for computers were card readers. The now almost obsolete IBM Hollerith card held exactly eighty characters. The first monitors were used as an adjunct to the card readers; the most obvious format was a width of eighty characters (one card per line). . . .
>
> As for the number of lines, I am sure that this was chosen for hardware reasons. However, the twenty-four-line standard is a multiple of eight, as is eighty, and computer people *love* eights.

In When Did Wild Poodles Roam the Earth? *we quoted an official of the American Banking Association who stated that the little pieces of white paper attached to the bottom of our checks were inserted when a typist made a mistake, and that these MICR numbers cannot be erased. The answer was correct as far as it goes, but reader Jeff Reese of Mosinee, Wisconsin, works for a company that sells a solvent that does erase those little numbers on the bottom of the check. He also adds that some banks use small stickers in lieu of a strip that runs along the bottom of the entire check.*

We told you in Poodles *that many drivers in cold climates wire cardboard to their grills to keep cold air from entering the engine. But one reader, Bruce Hyman of Short Hills, New Jersey, added some "lore" on the subject:*

Years ago, cars did not have thermostats in the radiator system, and the only way to keep the coolant warm enough (an oxymoron, of sorts) was to restrict the air flow to the radiator. Early cars sometimes had a pullchain from the radiator grill (which at that time was a set of venetian blind–type slats) into the passenger compartment, letting the driver control the amount of air to the radiator, and hence the engine water temperature.

At this point, these [makeshift equivalents of] "radiator blinds" are totally useless as long as the car has a functioning 180 degree Fahrenheit or hotter thermostat, because the thermostat controls the water temperature. Thermostats are placed in a housing with three hose connections: from the engine; to the radiator; and back to the engine. If the radiator gets too much cooling air, the thermostat simply closes down, and the water recirculates back to the engine through the "bypass" hose.

In When Do Fish Sleep? *we analyzed why most cameras are black. Freelance photographer Grace B. Weinstein of Los Angeles, California, adds two more reasons:*

Most cameras are black, with chrome trim, the chrome trim giving it a "classier" appearance. As a camera with chrome trim gets older and used more often, the chrome tends to wear off, while the black part stays black. It would take a pretty penny to have the worn chrome rechromed, but the black part can be refinished easily with matte paint. Another reason for a camera being all black is that chrome parts would catch the light and bring attention to the camera. A photographer who is working under cover would not like to bring attention to his camera because of the reflective catch lights in the chrome, which would act as a mirror that catches the light.

But not all our readers were concerned with technological Imponderables. Scott McDougall of Boise, Idaho, wanted to add another reason why dance studios are so often located on the second floor, a weighty question we tangled with in Do Penguins Have Knees?

As you noted in your answer, dance studios are most often located in commercial buildings, which usually have a concrete floor on the main level. Concrete is hard and nonresilient. Older commercial buildings of the type that typically house dance studios will have wood joists supporting a wood or wood composite floor beginning on the second level; this type of floor is gentler on the legs and provides more spring. That's the reason given when a dance company once rented space from me.

Speaking of movement, we received a long letter from Rick Ballard of East Lansing, Michigan, who says that he logs about 35,000 miles per year. He wanted to add to our discussion of why traffic on highways tends to clump together in bunches, which we discussed in Why Do Dogs Have Wet Noses?

> You correctly point out that bunching behavior on interstates is partly explained by slower cars in the passing lane, but you don't offer any reasons why they stay there. Several factors in rural interstate bunching need to be mentioned:
>
> - Once you have a bunch, it tends to stay a bunch because passing is physically more difficult for the faster cars behind. Some drivers would rather tailgate than pass. Truck convoys are the prime example. They effectively change two lanes into one, and inevitably increase bunching. The faster cars have to be in the passing lane longer just to get around. . . .
> - Cruise control is a mixed factor. It definitely contributes to extended time in the passing lane. Many cars will c-r-e-e-p around the slower car because they are unwilling to speed up slightly to complete their pass. Likewise, they are hesitant to go back into the right lane when passing a spaced string of traffic, because they may have to tap the brakes to wait for a faster car before they can pass the next car ahead. So instead, they continue passing a 63-mph string of traffic at 64 mph, blissfully ignoring the bunch growing behind them. But cruise control may also reduce bunching, by creating a disincentive to tailgate.
> - Drivers with radar detectors, and other fast drivers, contribute to bunching. Cars will frequently speed up to tag behind faster cars (especially with radar detectors) because they assume that if there is a speed trap, the driver ahead will either slow down or will be the one ticketed.

The psychology of drivers is an elusive one, indeed, but few humans are as fragile as the expectant father. Sandra Stout of Colonial Heights,

Virginia, felt that we missed the boat in our discussion in When Did Poodles Roam the Earth? *of why you see folks boiling water during home deliveries in movies and television.*

> Your answer was way off the mark. It has nothing to do with sterilizing things. I noticed that both of the experts you quoted were males, and you being male, the "sterilization theory" made sense. That is because the true purpose of the water boiling is for your gender.
>
> When a baby is delivered at home, the husband, under normal conditions, becomes concerned about the amount of pain his wife is in and wants to help as much as possible. What he ends up doing is getting underfoot. So the midwife or female in charge will send the male, or in some cases the other children in the family, to boil some water so that he is out of the way but feels as if he is being helpful. That is why you never see the water—it never is used for anything.

More male bashing, heh. Doesn't anyone love us? Well, we guess Karen Friend of Baltimore, Maryland, loves us, sort of:

> I'm a big fan of yours. Although I have no practical knowledge, I've read all your books and can consequently impress hordes of people at cocktail parties.

Gee, Karen, there's nothing you could possibly say that would make us prouder.

Speaking of pride, we have to eat ours, for we made a misstatement in When Did Wild Poodles Roam the Earth? *We stated that one of the reasons why horses might not vomit is because they must eat at a virtually nonstop pace to derive proper nutrition. We heard from veterinarian Robert E. Habel, one of the foremost ungulate specialists on the planet:*

> The horse does not have to keep its stomach full; it has a capacious intestine and does not have to eat continuously, as thousands of trail riders could tell you.
>
> My theory is that the main difference between the vomiters and the nonvomiters (other than alcohol consumption) may be in the brain, where the vomiting reaction center is located. Dogs and people may have evolved this protective mechanism because they swallow lots of nasty things (as you point out) that need to be vomited. Cows and horses don't.

And while we are eating crow (but not vomiting it), may we apologize for a silly error in Poodles? *In an offhand analogy in our discussion of why people don't wear hats as often as they used to, we contrasted the impact of John F. Kennedy's hatless inaugural with the popularity of the undershirt after Clark Gable flaunted one in* It Happened One Night. *We meant, of course, to say that Gable appeared without an undershirt and that sales of the apparel quickly plummeted (or so the legend goes). We were surprised that only two readers, Arnold Hecht of Greensboro, North Carolina, and Charles Raphael of Montreal, Quebec, wrote to nail us on what may be our most ridiculous mistake since we transposed the meanings of "diameter" and "circumference" in our first book,* Imponderables.

Hmmm. Now we seem to be self-bashing. Let's escape to the comfort of hearing from some readers theorizing about perennial Imponderables. You still seem to want to talk about why there are dents on the top of cowboy hats, which we first discussed in Do Penguins Have Knees? *Daniel Coppersmith of Mishawaka, Indiana, weighs in with an interesting theory:*

> I have always believed that the dent was to allow rainwater to drain off the hat. If the hat were flat, water would stand on top of the hat and increase the chances of the water soaking in and leaking on the head of the cowboy. The front of a properly worn cowboy hat has a lip on the front brim. This keeps the water rolling off the top of the hat from splashing onto the face of the wearer.

We always thought the brim was more important for keeping the sun out of the eyes, and there's no particular reason why a cowboy hat couldn't have a rounded top. But since you didn't bash us, we're hardly going to insult you. Nor will we bad-mouth John A. Steer of Durham, North Carolina, whose surname lends him special qualification to theorize about cowboy hats:

> Here is another approach: How about convenience? In the days when cowboy hats were designed nearly all men, even cowboys, behaved as gentlemen should and tipped their hats on meeting and greeting ladies.

In the East at that time, men wore derbies, homburgs, and even top hats. These all have stiff brims; thus, without a dent to grab hold of, they could still easily tip their hats. The cowboys' rough life required a soft hat that could be knocked back into shape when required—for some that may have been often. The dent was their hat handle.

We're almost through the Letters section, but there seems to be something missing. What could it possibly be? Oh, of course. A discussion of boots on the top of ranchers' fenceposts. It's with a slowly trickling tear down the side of one cheek that we announce that this is the first year since the publication of Why Do Clocks Run Clockwise? *that we received no new theories about the resons for their appearance. But at least Robert M. Brown of Bellevue, Washington, provided us with an exciting new sighting. Robert sent us a clip from his local paper, the* Journal American, *in which reporter Liz Enbysk profiles Warren Oltmann, whose boots are displayed on Highway 203 south of Carnation, Washington:*

> "If I can make one person smile per day I figure I've done something good," Oltmann figures. Besides, what else would you do with seven pairs of worn-out boots?

We feel the same way. If we can make one person smile per day, we're happy. Strike that. We'll settle for one living thing and/or one feline per day. We received this note from Catherine Greene, a ninth-grader from Silver Spring, Maryland:

> I adore your books, and so does my cat, Frances. She sits in my lap and pretends to read your books while I read them!

Better this way, Catherine, than if Frances read the books and you pretended to read them.

Having contended already in this book with the neverending debate about why we find one shoe on the side of the road so often, we are loath to open up the following can of worms. Still, we couldn't resist sharing this related conundrum, sent to us by Charlene Ingulfsen of Asheville, North Carolina:

Why do I so often see loops of audio cassette tape beside the road? I've never lost a tape out the window of my car, but maybe that's the way others dispose of their tapes.

Something else I've noticed—I recently moved to North Carolina from Oklahoma. Loops of tape by the roadside aren't as common here as they were in Oklahoma and Texas. I wonder why. But I spent several months in Norway, and I often saw tape on the roadside in that lovely country as well.

Off the top of our head, Charlene, wouldn't it seem most likely that you are seeing the remnants of cassettes that were jammed in auto cassette decks? But do you expect us to know about the regional variations in audio cassette side-of-the-road dumping? Golly, do you think we are everywhere? Well, come to think of it, at least one reader, Ken Giesbers of Seattle, Washington, does think we're everywhere:

You live in New York City. HarperCollins Publishers are based in New York City. You conducted a survey of Philadelphia bakeries in response to the Imponderable in *When Did Wild Poodles Roam the Earth?* about seven layer cakes. So why do we send our Imponderables to Los Angeles? Are you omnipresent?

We've been reluctant to admit it until know, Ken, but yes indeed, we are omnipresent. It's hard work, being omnipresent, but we will stop at nothing to gather knowledge in our ever-vigilant fight to stamp out Imponderability. As a matter of fact, we have to sign off right now so that we can fight the good fight.

So have a safe and happy year, readers. And please behave: We can check up on you. Don't you remember? We're omnipresent.

Acknowledgments

Change is a constant, but the support and enthusiasm of the readers of Imponderables books is one of the few things I can always count on. Your letters are the lifeblood of this book, not only because you offer Imponderables and Frustables solutions but because your criticisms let me know where I've gone wrong, and your praise inspires me to keep working.

In the past year, I've tried to reply to your letters in a speedier fashion; most of the time, I have succeeded. Please be patient if it takes me a while to get back to you. Particularly when I'm on the road publicizing a book or in the throes of writing a new manuscript, I may have to slow down the correspondence. But rest assured I read every word of every letter I receive and still answer all mail that comes with a self-addressed stamped envelope.

It is with bittersweetness that I acknowledge my editor, Rick Kot, for all of his help over the last six years. Rick is leaving behind a trail of friends and admirers as he departs HarperCollins, and I'm at the top of that long list.

Rick's assistant, Sheila Gilooly, a gifted editor in her own right, has also flown the coop to pursue bigger and better things. Thank you, Sheila, for making my work not only easier but more fun.

Thankfully, all my other "main squeezes" at HarperCollins are still around to make it a wonderful place for an author. Thanks to Craig Herman for guiding my publicity for the past five years and to Wende Gozan for her indefatigable enthusiasm while booking my tour for *Poodles* and this book (this is her *Headache* as well as mine).

Every year, it seems the production schedule gets more hectic. Special thanks to the team that help make my prose coherent and the package attractive: Kim Lewis, Maureen Clark, Janet Byrne, Karen Malley, and Suzanne Noli.

Even the Grand Pooh Bahs at HarperCollins have been consistently gracious and supportive. Thanks to Bill Shinker, Roz Barrow, Brenda Marsh, Pat Jonas, Zeb Burgess, Karen Mender, Steve Magnuson, Robert Jones, Joe Montebello, Susan Moldow, Clinton Morris, and Steven Sorrentino. Connie Levinson and Mark Landau, and the entire special markets department, have been a constant source of creative ideas and friendship. And we can't even talk about the achievements of the Academic and Library Promotion Department (Joan Urban, Diane Burrowes, Virginia Stanley, and Sean Dugan)—after all, this is a family book. And a salute to all my friends in the publicity, sales, and Harper Audio divisions.

My agent, Jim Trupin, is like my Boswell—Tom Boswell, that is. Sorry, I couldn't resist a cheap joke, which is why Jim and I get along so well—I'm willing to listen to his clunkers, too. If Jim's my Boswell, then Liz Trupin's my Marion Ross, and my days have been happier since I've known her.

Kassie Schwan is one of the few illustrators I know who can make aspirin look interesting. Come to think of it, she's managed to make nine of my books look interesting. Thanks for your wonderful work.

Special gratitude is owed to friends in the publishing business who allow me to natter on about my problems and reciprocate by offering warm shoulders and sage advice. Thank you, Mark Kohut, Susie Russenberger, Barbara Rittenhouse, James Gleick, and all of my Prodigy pals, especially the Housewife Writers.

Let us now praise those whose research proved invaluable. Thanks to Sherry Spitzer, Judith Dahlman, Honor Mosher. David Schisgall, and Chris McCann for making my work so much easier.

Thanks to all my friends and family, who actually managed to tolerate me for another year: Jesus Arias; Michael Barson; Sherry Barson; Rajat Basu; Ruth Basu; Barbara Bayone; Jeff Bayone; Jean Behrend; Marty Bergen; Brenda Berkman; Cathy Berkman; Sharyn Bishop; Andrew Blees; Carri Blees; Christopher Blees; Jon Blees; Bowling Green State University's Popular Culture Department; Jerry Braithwaite; Annette Brown; Arvin Brown; Herman Brown; Ernie Capobianco; Joann Carney; Lizzie Carnie; Susie Carnie; Janice Carr; Lapt Chan; Mary Clifford; Don Cline; Dori Cohen; Alvin Cooperman;

Audrey Cooperman; Marilyn Cooperman; Judith Dahlman; Paul Dahlman; Shelly de Satnick; Charlie Doherty; Laurel Doherty; Joyce Ebert; Pam Elam; Steve Feinberg; Fred Feldman; Gilda Feldman; Michael Feldman; Phil Feldman; Ron Felton; Kris Fister; Mary Flannery; Linda Frank; Elizabeth Frenchman; Michele Gallery; Chris Geist; Jean Geist; Bonnie Gellas; Richard Gertner; Amy Glass; Bea Gordon; Dan Gordon; Emma Gordon; Ken Gordon; Judy Goulding; Chris Graves; Christal Henner; Lorin Henner; Marilu Henner; Melodie Henner; David Hennes; Paula Hennes; Sheila Hennes; Sophie Hennes; Mitchell Hofing; Steve Hofman; Bill Hohauser; The Housewife Writers; Uday Ivatury; Terry Johnson; Sarah Jones; Allen Kahn; Mitch Kahn; Joel Kaplan; Dimi Karras; Steve Kaufman; Robin Kay; Stewart Kellerman; Eileen Kelly; Harvey Kleinman; Claire Labine; Randy Ladenheim-Gil; Julie Lasher; Debbie Leitner; Marilyn Levin; Vicky Levy; Rob Lieberman; Jared Lilienstein; Pon Hwa Lin; Adam Lupu; Patti Magee; Rusty Magee; everybody at the Manhattan Bridge Club; Phil Martin; Chris McCann; Jeff McQuain; Julie Mears; Phil Mears; Roberta Melendy; Naz Miah; Carol Miller; Honor Mosher; Barbara Musgrave; Phil Neel; Steve Nellisen; Craig Nelson; The Night Owl Chat gang; Millie North; Milt North; Charlie Nurse; Debbie Nye; Tom O'Brien; Pat O'Connor; Terry Oleske; Joanna Parker; Jeannie Perkins; Merrill Perlman; Joan Pirkle; Larry Prussin; Marlys Ray; Joe Rowley; Rose Reiter; Brian Rose; Lorraine Rose; Paul Rosenbaum; Carol Rostad; Tim Rostad; Leslie Rugg; Tom Rugg; Gary Saunders; Joan Saunders; Mike Saunders; Norm Saunders; Laura Schisgall; Cindy Shaha; Pat Sheinwold; Aaron Silverstein; Kathy Smith; Kurtwood Smith; Susan Sherman Smith; Chris Soule; Sherry Spitzer; Stan Sterenberg; Kat Stranger; Anne Swanson; Ed Swanson; Mike Szala; Jim Teuscher; Josephine Teuscher; Laura Tolkow; Albert Tom; Matddy Tyree; Alex Varghese; Carol Vellucci; Hattie Washington; Ron Weinstock; Roy Welland; Dennis Whelan; Devin Whelan; Heide Whelan; Lara Whelan; Jon White; Ann Whitney; Carol Williams; Maggie Wittenburg; Karen Wooldridge; Maureen Wylie; Charlotte Zdrok; Vladimir Zdrok; and Debbie Zuckerberg.

Because Imponderables are questions with answers not easily found in books, we rely on the wisdom of experts in every imaginable field. Although we contacted close to 1,500 corporations, universities, foundations, professors, associations, and miscellaneous experts on everything from acrylic to yogurt, we thank here only the sources who led directly to the answers to the Imponderables in this book:

Morris Adams, Thomas Built Buses; Tony Aiello, California Beef Council; Bob Alter, Polaroid Corporation; Harry Amdur, American Photographic Historical Society; John Anderson, Borden's Glue; Carl Andrews, Hershey Foods Corporation; Joe Andrews, Pillsbury Brands.

Bill Baker, Recreational Vehicle Industry Association; Joseph P. Bark; Kim Barley, White Castle; Andrea Bean, ERIC; Kathie Bellamy, Baskin-Robbins; Elizabeth Berrio, Immigration and Naturalization Service; Susan Berry; Melissa Bertelsen, First Data Resources, Inc.; Jim Boldt, Great Northern Corporation; Al Brock, WKLX; Don Buckley, National Ice Cream Retailers Association; Robert Burnham, *Astronomy*; Russell Burns, American Racing Pigeon Union; Steve Busits, American Homing Pigeons Fanciers.

John Cahill, Fine Art Photography; Frank Calandra, Photographic Historical Society; Rick Campbell, NCAA; Jeff Carpenter, Washington State Superintendent of Physical Education and Health Education; Bob Carroll, Pro Football Researchers Association; Ellen Carson, Empire Berol USA; David Cerull, Fire Collectors Club; Lynn Chen, American Numismatic Association; Officer Claggett, Richmond, Virginia Police Department; Annette Clark, Scottish Heritage Association; Calvin Clemons, National Association of Writing Instruments; Linda W. Coleman, Bureau of Engraving and Printing; Mark T. Conroy, National Fire Protective Association; John Cook, Georgia Dermatology and Skin Cancer; Tony Crawford, Dallas Police Department; Todd Culver, Cornell University Laboratory of Ornithology; Kimberly J. Cutchins, National Peanut Council.

Sandy Davenport, International Jelly and Preserve Association; Harold Davis, FDA; Thomas Deen, Transportation Research Board; Georgeanne Del Canto, Brooklyn, New York Board of Education; Arthur Douglas, Lowell Corporation; Joe Doyle; Bill Dwyer, Federal Signal Corporation.

Merle Ellis; Kay Engelhardt, American Egg Board; Stuart Ensor, National Live Stock & Meat Board; Marcus Evans, GACCP.

Darryl Felder, University of Louisiana, Lafayette; Fred Feldman; Karen Finkel, National School Transportation Association; Sheila Fitzgerald, Michigan State University; Dan Flory, Cincinnati College of Mortuary Science; William Frank, Aluminum Company of America; Meryl Friedman, Mattel Toys.

Samuel R. Gammon, American Historical Society; Marsha R. Gardner, Hershey Foods; Brenda Gatling, United States Mint; Jay Gilbert; Joan Godfrey, Alpines International; Emmanuel Goldman, *Curriculum Review*; Ailette Gomex, California Pistachio Commission; Michael Goodwin, International Paper; Stanley Gordon, Bridge Division, Federal Highway Administration; Lee Grant, Agricultural Extension Service, University of Maryland; Tina Steger Gratz, American Dental Association; Robert Gray, Old Dominion Box Company; Gene Gregory, United Egg Producers.

Officer Hallock, Salt Lake City Police Department; Herb Hammond, New York Rangers; Paul Hand, Atlantic Dairy Cooperative; Darryl Hansen, Entomological Society of America; Charles E. Hanson, Museum Association of the American Frontier; Pat Harmon, College Football Hall of Fame; Ruth Harmon, Miller Brewing Company; Megan Haugood, California Table Grape Commission; John Bell Henneman, ICHRPI; Martin Henry, National Fire Protection Association; David Hensing, AASHTO; Billy Higgins, AASHTO; Janet Hinshaw, Wilson Ornithological Society; Bruce Hisley, National Fire Academy; Francis Hole; David Hoover; Ike House; John Howland, American Goat Society; Richard Hudnut, Builders Hardware Manufacturers Association; John Husinak, Middlebury College; W. Ray Hyde.

Peter Ihrke, American Academy of Veterinary Dermatology.

Simon S. Jackel; Sally Jameson, Braille Institute; Alison Johnson, Delta Airlines; Michele Waxman Johnson, Institute of Transportation Engineers.

Maggie Kannan, Department of Photographs, Metropolitan Museum of Art; Kelly Karr, American Meat Institute; Phil Katz, Beer Institute; V. Herbert Kaufman, SAE International; Carol Keis, Hanna-Barbera Productions; Glenda Kelley; Lynn Kimsey, Bohart Museum of Entomology; Anthony L. Kiorpes, University of Wisconsin-Madison School of Veterinary Medicine; Ben Klein; Phil Klein; Robin Klein, Pork Information Bureau; Don Koehler, Georgia Peanut Commission;

Richard Kramer, National Pest Control Association; Sharon Kulak, National Live Stock & Meat Board.

Catherine R. Lambrechts, Dean Witter Financial Services Group; Richard Landesman, University of Vermont; Michael de L. Landon, University of Mississippi; Christopher Landry, New Orleans Police Department; Gerald Lange, Alliance for Contemporary Book Arts; Michael Lauria; Kenneth Leavell, Du Pont; Thomas A. Lehmann, American Institute of Baking; Rafael Lemaitre, Museum of Natural History; Joe Lesniak, Door and Hardware Manufacturing Institute; Gene Lester, American Society of Camera Collectors; Barbara Linton, National Audubon Society; Jerome Z. Litt; Theodore Lustig, West Virginia University; Denny Lynch, Wendy's.

Steward MacBreachan; Alan MacRobert, *Sky & Telescope*; Judy Maire, National Association of School Nursing Consultants; Ed Marks, *The Ice Screamer*, Bob Martin, Scottish Tartan Society; Nancy Martin, Vermont Institute of Natural Science; John T. McCabe, Master Brewers of the Americas; Roy McJunkin, California Museum of Photography; Pat McKelvey, Wine Institute; Lisa McKendall, Mattel Toys; Carmen Miller, Helen of Troy Corporation; Patricia Milroy, McDonald's; Joseph Mitchell, University of Richmond; Randy Morgan, Cincinnati Insectarium; Rose Motzko, Walt Disney; Michelle Mueckenhoff, Dairy Council of Wisconsin.

Jean Naras, American Dairy Association; National Collegiate Athletic Association; Kevin Newsome, Newsome's Studio of Photography; James Nolan, Fibre Box Association; Human Numan, Z-100.

Richard O'Brien; Barnet Orenstein; Alberta Orr, American Foundation for the Blind; Betsey Owens, Virginia-Carolina Peanut Promotion.

Richard Palin, Loctite Corporation; Tonya Parravano, National Live Stock & Meat Board; Peanut Factory; Joan Peyrebrune, Institute of Transportation Engineers; Bob Phillips, American Racing Pigeon Union; Planters Corporation; Jay Poster, *Agronomy News*; Mary Potter, Mead Corporation; Bill Powell, Booth Memorial Medical Center; Gary Priest, San Diego Zoo.

Howard Raether; T.E. Reed, American Rabbit Breeders Association; Donna Reed, National Pasta Association; Carl Reichenbach,

Dixon Ticonderoga Company; Viva Reinhardt; Revlon, Inc.; John-Paul Richluso, American Association for State and Local History; Robin Roche, San Francisco Zoological Society; John R. Rodenburg, Federated Funeral Directors of America; Janet Rodowca, Fort Howard; Robert R. Rofen, Aquatic Research Institute; Grant Romer, George Eastman House; Officer Romero, Los Angeles Police Department; Mary H. Ross, Virginia Polytechnic Institute; Kate Ruddon, American College of Obstetricians and Gynecologists; Tonda Rush, National Newspaper Association; John Rutkauskas, American Society for Geriatric Dentistry.

Walter Sanders, Diners Club International; Richard Santore, Associated Funeral Directors Service International; Leslie Saul, San Francisco Zoological Society; Ray Schmidt; Robert Schmidt, North American Native Fishes Association; Paul Schofield, Discover Credit Corporation; Jim Schreier, Council on America's Military Past; Don Schwartz, Children's Hospital; Ira Schwartz, New York State Regents Advisory Committee; Norman Scott, Society for the Study of Amphibians and Reptiles; Samuel Selden; Arlene Sheffield, New York State Education Department; Ron Siebel, J.E. Siebel Sons' Company; John Simmons, American Society of Ichthyologists and Herpetologists; Charles Simpson, Texas A&M; Harry Skinner, Traffic Engineering Division, Federal Highway Administration; Philip and Shirley Smith; Bob Smith, Cytec; Bruce V. Snow; Stephen Snyder, Book Manufacturers Institute; Daniel Solomon, E&J Gallo Winery; Jeff Solomon-Hess, *Recycling Today*; Kent Sorrells, Altadena Certified Dairy; Dennis Stacey; Lynne Strang, Distilled Council of the United States; Robert Strawn, RVDA of North America; Joe Struble, George Eastman House.

Anne Tantum, American Association of Meat Processors; Thad W. Tate, College of William & Mary in Virginia; Farook Taufiq, Prince Company; Karl Torjussen, Westvaco Corporation.

United Airlines; USDA—Agricultural Research Service.

Pamela Van Hine, American College of Obstetrics and Gynecologists; Vanner Weldon, Inc.; Richard Van Iderstine, National Highway Traffic Safety Administration; Matthew J. Vellucci, Distilled Spirits Council of the United States; Ralph E. Venk, Photographic

Society of America; Rhoda Virgil-Madison, Detroit Police Department.

Larry Wallace, Monsanto; Roscoe Wallace, Monsanto; Steve Warren, MOR Media; James Weidman, Welch Foods, Inc.; Rebecca Weller, Fuller Brush Company; Vicki Wells, Miles, Inc.; Thomas Werner, New York State Department of Transportation; E.J. Westerman, Kaiser Aluminum and Chemical Corporation; Ron Whitfield, Marine World Africa USA; Steve Whitmire, Jim Henson Productions; Vann Wilber, American Automobile Manufacturers Association; Richard Wisniewski, University of Tennessee, Knoxville; Dori Wofford, Levi Strauss & Company; Wilton Wong, Images Unlimited Photography; Danny Wright; Jim Wright, New York State Department of Transportation.

Bruce and Janice-Carol Yasgur; Karen Yoder, Entomological Society of America.

And to the many sources who, for whatever reason, preferred to remain anonymous, our sincere thanks.

Index

Help!

We hate to end the book on a downbeat note, but we have to admit one dread fact: Imponderability is not yet smitten. Let's stamp it out.

"How?" you ask. Send us letters with your Imponderables, answers to Frustables, gushes of praise, and even your condemnations and corrections.

Join your inspired comrades and become a part of the wonderful world of *Imponderables*. If you are the first person to submit an Imponderable we use in the next volume, we'll send you a complimentary copy, along with an acknowledgment in the book.

Although we accept "snail mail," we strongly encourage you to e-mail us if possible. Because of the volume of mail, we can't always provide a personal response to every letter, but we'll try—a self-addressed stamped envelope doesn't hurt. We're much better with answering e-mail, although we fall far behind sometimes when work beckons.

Come visit us online at the Imponderables website, where you can pose Imponderables, read our blog, and find out what's happening at Imponderables Central. Send your correspondence, along with your name, address, and (optional) phone number to:

feldman@imponderables.com
www.imponderables.com

Imponderables
P.O. Box 116
Planetarium Station
New York, NY 10024-0116

YOU REALLY DON'T NEED US FOR THAT HEADACHE! FOR RELIEF, TAKE DOZENS OF READER LETTERS!
GREAT!
IMPONDERABLES
WHERE'S
WHY'S WH
WHAT'S WHAT
WHEN'S WHEN
HOW'S HO

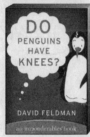

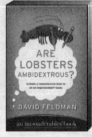

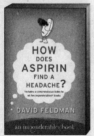

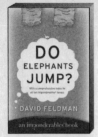

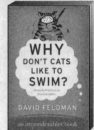

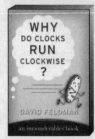

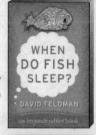